The Digital Innovation Race

Cecilia Rikap · Bengt-Åke Lundvall

The Digital Innovation Race

Conceptualizing the Emerging New World Order

Cecilia Rikap
City, University of London
London, UK

CONICET
Buenos Aires, Argentina

Bengt-Åke Lundvall
Aalborg University
Copenhagen, Denmark

ISBN 978-3-030-89442-9   ISBN 978-3-030-89443-6   (eBook)
https://doi.org/10.1007/978-3-030-89443-6

This Palgrave Macmillan imprint is published by the registered company Springer Nature Switzerland AG
The registered company address is: Gewerbestrasse 11, 6330 Cham, Switzerland

PREFACE

The collaboration leading to this book began almost two years ago and it started with a meeting in cyberspace between two scholars with quite different backgrounds. Bengt-Åke, a senior male European working within economics and management of innovation and knowledge and Cecilia, a young woman scholar from Latin America working on corporate power in the era of intellectual monopoly capitalism. This is reflected in the analytical perspectives in the book where insights from innovation studies are combined with political economy perspectives.

What we have in common is a deep concern regarding current trends in world development. While the enormous resources invested in science and advanced technologies have created the potential to establish an egalitarian and democratic world that could put a stop to global warming as well as global pandemics, at present, these opportunities are squandered.

This gap, between what could be done to create a sustainable world and what we actually witness, reflects, as we see it, two fundamental characteristics of world governance: the transformation of knowledge into private property and techno-nationalism. With fundamental changes in these two dimensions, there are grounds for optimism.

As compared to tangible resources, knowledge has unique characteristics. It is not a scarce resource—the more you use it the more you get. This is certainly true for experience-based knowledge, where people and organizations learn from what they do. In the current era of artificial intelligence, in particular of machine learning, a similar cumulative mechanism is at work when algorithms are trained through input streams of data.

Within this general perspective, we have focused on digitalization as a second phase of the information and communication technology revolution. Artificial intelligence and digitalization have been hyped for decades, but it was not until recently that this combination has shown an enormous potential both to stimulate economic development and to tackle global challenges related to health and the environment. However, the national and global institutional set-ups hamper the realization of this potential.

As the control of these strategic technologies gets concentrated in a handful of private corporations located in two superpowers increasingly engaged in technological warfare among themselves, there is less room for directing these powerful technologies to address global challenges. Current regulatory regimes and modes of governance are inadequate when it comes to handle this new era of globalized intellectual monopoly capitalism. This is true for national and regional regimes and not least for the weakly developed and contested global regime characterized by a transition from a US-dominated world order to a more multipolar but also more disorderly world.

Our general perspective has taken on new significance in the wake of the Covid crisis. While the speed of developing new vaccines illustrates the power of science, the access to vaccines has been absurdly uneven, with rich countries hoarding vaccines, leaving the population in poor countries with very limited access. Building on research financed by taxpayers, big pharma companies have enriched their shareholders by selling vaccines at prices reflecting the extraction of substantial rents. Even within the rich countries, there is now growing concern that capricious private actors and nationalist strategies determine their citizens' access to jabs.

Furthermore, big pharma companies have expanded their intellectual monopolies since the pandemic started. The enormous amassment of financial resources and intangible assets by digital tech giants documented in this book got another boost from the Covid-19 global crisis. Restrictions on face-to-face interaction resulted in rapid growth in digital services and, as a result, these intellectual monopolies reinforced their worldwide dominance in terms of markets as well as technologies. So, while it took us some time to finalize the book, the delay did not make it lose its relevance.

Overall, this book presents rich and detailed empirical data on digital tech giants, the role of artificial intelligence and the techno-nationalist strategies of the US and China. In order to give meaning to these data, we introduce them within theoretical framings with roots in political economy and innovation studies. The intention is to contribute to raising awareness of the scale of the changes taking place and, ultimately, stimulate the formation of programs of action. In order to do so, we offer theoretical conceptualizations more adequate than what students can find in standard economic textbooks.

London, UK   Cecilia Rikap
Copenhagen, Denmark   Bengt-Åke Lundvall

CONTENTS

LIST OF FIGURES

LIST OF TABLES

CHAPTER 1

A Global Race Between Giant Corporations and Nation States

Abstract This book conceptualizes the global innovation race between nation states and corporations in the second phase of a technological revolution characterized by Artificial Intelligence and explosive growth in digital services. Each chapter elaborates on an existing or new conceptual framing to analyse empirical dimensions of the economics and politics of innovation in the context of digital capitalism. In this introductory chapter, we set the scene by presenting the main actors and highlight the crucial role of Artificial Intelligence. We end it by introducing the structure of the book.

Keywords Innovation race · Artificial intelligence · US–China conflict · Tech giants · Globalization

1.1 INTRODUCTION

The global innovation race refers both to the technology war between the United States (US) and China and to the struggle between giant corporations from these two countries to establish world dominance in artificial intelligence (AI), in particular machine learning and related technologies. The recent acceleration in the development and use of digital technologies, spanning from 5G communication systems to cloud services and

2 C. RIKAP AND B.-Å. LUNDVALL

AI, is transforming the world order. It gives rise to extreme degrees of concentration of capital, and it offers a window of opportunity for China to challenge the US as a hegemonic power.

The US and Chinese governments increasingly engage in techno-nationalist strategies. They use science, technology and innovation policies to promote strategic (digital) technologies. They use protectionist measures to reduce rivals' access to critical resources in order to slow down their technological development. They engage in shaping international intellectual property regimes to promote national interest and standard setting. Concerning the latter, they make diplomatic efforts and design projects in third countries to promote the diffusion and use of their own technology standards. In Chapters 6 and 7 we analyse in more depth how the state in China and the US engages in the AI innovation race.

The other main participants in the race are a handful of giant firms from these two countries: Google, Amazon, Facebook, Apple and Microsoft (GAFAM) in the US and Baidu, Alibaba, Tencent and Huawei (BATH) in China. They operate production and innovation globally, offshore earnings, minimize paid corporate taxes and they replace states in the definition of norms and regulations of the digital economy. In Chapter 2, we focus on how they operate as intellectual monopolies harvesting rents from their global operations. In Chapter 3, we see how each of them builds and profits from its own global corporate innovation system.

The twenty-first-century global innovation race is shaping the world's future. It has an impact on world economic growth and on how economic gains and losses will be distributed between and within economies. In Chapters 4 and 5 we analyse the technologies emerging from this race and we show how they are applied to a wide set of fundamental economic and social activities, hereby affecting the future of work, health and culture, including human rights and privacy as well as the climate and other ecological challenges. The more intense rivalry between two great powers raises fundamental issues of war and peace. To explore what lies behind this innovation race is, therefore, of great urgency and a prerequisite for finding ways to cope with its consequences.

To analyse this historically unique constellation, where technology has become the central subject of geopolitical conflict and giant corporations challenge the sovereignty of nation states, requires alternatives to the conceptual frameworks that still dominate textbooks in standard economics. In this book, we combine insights from political economy and innovation studies and propose alternative theoretical concepts at three different levels: the corporation, technology and the state.

In Chapter 2, we introduce the concept *intellectual monopoly*, offering a different perspective of economic power, and in Chapter 3 we elaborate on the concept, *corporate innovation system*, offering a new perspective on inter-firm relationships and innovation. To capture the technological specificity and evolution of AI we analyse it as a *technological innovation system* in Chapter 4. Moreover, to capture its evolution we refer to the concepts *technological paradigm and trajectory* in Chapter 5 focusing on Amazon and Microsoft's technological convergence. To understand the roots of the technology war between the US and China, we refer to the literature on *national innovation systems* and *catching-up* in Chapter 6 and to the concepts *techno-nationalism and techno-globalism* in Chapter 7.

In the rest of this chapter, we present the main actors and highlight the crucial role of AI in contemporary capitalism. The chapter ends with an introduction of the structure of the book.

1.2   Why Technology War?

Why have the US and the Chinese governments moved in the direction of techno-nationalism, and why has the US declared technological war on China? Answering these questions requires an analysis of the specific character of the emerging new technologies and of the current geopolitical context. Nonetheless, we can draw important lessons from world history. While the specific constellation of AI as a transformative technology and the US and China as the major national actors is unique, there are historical precedents of how a transformative technology plays an important role in upsetting the ruling world order. In Chapter 6, we refer to the work by Christopher Freeman and Carlota Perez on how technological revolutions open up windows of opportunity for catching-up and challenge world leadership.

1.2.1   Artificial Intelligence and the Critical Role of Big Data

In order to understand why leadership in a specific technology has emerged as a major issue in geopolitics, confronting the US with China, it is necessary to look closer at the character of the technology. In the following sections, we will point to the importance of cumulative effects and scale economies, emanating from the combination of major advances in AI and supporting technologies related to telecommunication and supercomputers in combination with major organizational innovation, including privately managed platforms and cloud computing.

At the world level, this has resulted in an extreme concentration of capabilities in the biggest and most data-rich countries, the US and China. The cumulative character of technological advance in AI, where the inflow of data upgrades the technology itself, makes leaders in both countries fear that getting behind in the innovation race would lead to a downward spiral in terms of economic and political power.

In Chapters 4 and 5 we focus on AI as the most important of the emerging new technologies characterizing the second phase of the ICT revolution. AI refers to a set of related technologies that make it possible for computers to imitate human thinking and learning from experience. Following WIPO (2019), AI systems are learning systems, where machines get better at a previously human task with limited or no human intervention. Between 2011 and 2017, patent applications in AI grew by a factor of 6.5 per year—i.e. with staggering 650% per annum.

Within AI techniques, machine learning dominates. It can be defined as "an AI process that uses algorithms and statistical models to allow computers to make decisions without having to explicitly program it to perform the task" (World Intellectual Property Organization, 2019, p. 146). Machines learn to solve problems by processing big data, thus depending on access to permanent streams of data.[1]

The most popular machine learning approaches are deep learning and neural networks (World Intellectual Property Organization, 2019).[2] Examples of AI functional applications based on deep learning and neural

---

[1] Tech giants' dependence of access to big data makes them vulnerable to regulations that restrict their access. As we demonstrate in Chapter 5, they are aware of this risk and invest substantial resources in developing new forms of machine learning where they can reuse data or train algorithms with smaller volumes of data.

[2] Deep learning refers to hierarchically layering data or information. It builds hierarchies of concepts with different levels of genericity (think for instance of the Linnaean taxonomy). Neural networks try to emulate humans' interconnected neurons. It is an approach where an interconnected framework of functions processes multiple data inputs. Using both approaches together (building different deep learning layers of neural networks) leads to far and wide AI-learning systems where each layer's output is an input of the following layer.

network approaches are speech processing, recognition and synthesis, natural language processing and images and video segmentation.

Algorithms are at the very core of machine learning models. An algorithm is a set of rules or instructions to be followed for problem-solving operations or to perform a computation. Algorithms are used to process data in computing systems. The specificity of machine learning algorithms is that they evolve and become more "knowledgeable" on the basis of inflow and analysis of data. This is why scale is important for the development and use of AI. Machine learning requires big volumes of data.

Scholars have compared data to raw oil and cloud services and datacentres to oil refineries. But this parable is misleading in several respects. First, while oil is a rival good that can be used only once, data is non-rival—data points can be used again and again without losing their value. Second, in terms of impact on value creation, machine learning (the transformation and upgrading of the algorithm) is more important than the refined data. Sticking with the parable, the most important outcome is not the refined oil but a more productive refinery.

In the digital era, the issue about *excludability* of data is at the core of political and ethical concerns. Some data exist as knowledge commons to which everybody have access—examples are public databases and statistics, phone books and maps. Other data remain secret and/or private for different reasons. Sometimes, the privacy of data is protected by law. Individuals usually prefer to keep some data private rather than public (such as information on their health and on their political preferences). Firms rely on specific technological knowledge that they do not want to share with competitors. States protect military and diplomatic secrets, sometimes with death penalty for those breaking the rules, and central and commercial banks keep banking and broadly financial data secret. This is another reason why this field is unique in relation to other economic activities, such as the oil industry. Digital technologies raise issues about individual human rights, industrial espionage and national sovereignty, offering one explanation of why states gets so strongly engaged in this specific innovation race. Another, at least as important, reason is the economic potential of the new technologies.

1.2.2   Platforms, Cloud Computing and Economic Concentration

The recent acceleration in the development and use of AI reflects major organizational innovations such as the formation of digital platforms and cloud computing controlled by tech giants. The resulting concentration of intangible assets within a few corporations from the US and China fuels the conflict between the two countries.

The first phase of the ICT revolution, based on microelectronics and desktop computing, transformed working life. Widespread mobile communication gave new opportunities for citizens and enterprises all over the world to exchange information. When this first phase was at its early stage—around 1975–1985—there were exaggerated expectations on its short- and medium-term impact on employment and productivity. In retrospect, we can see that the quantitative changes took place more gradually than expected, while the long-term qualitative effects on work, economic organization and everyday life have been dramatic. This first phase lay the ground for the one we are currently in the midst of.

One major difference is that, while the first ICT phase gave room for decentralization, new entrepreneurial entrants and increased competition, the second phase is predominantly characterized by economies of scale and scope leading to extreme degrees of concentration and centralization of capital.

One important organizational innovation leading to industrial transformation and market concentration is the formation of "technology platforms" owned by private enterprises and characterized by unique mechanisms driving economies of scale and scope. Platforms generally involve three parties: The owner of the platform (Amazon's marketplace, Google's search engine, Facebook, Airbnb, Uber, etc.) and two types of *clients*, the service provider (which for gig economy platforms are unrecognized workers) and the service user.

Platforms are characterized by complex *network effects* both when it comes to providers and users. First, as the number of users increases, more users are encouraged to join—cf. Facebook. Second, the increased number of users makes the platform more attractive for providers (advertising firms in the case of Facebook). Likewise, as the number of providers increases, more users are attracted since they get access to a more diverse offer of goods and services (cf. Amazon). This implies that incumbents have an advantage that makes it difficult for newcomers to enter platform markets. The emergence of platforms contributes to the general tendency of economic concentration and centralization leading towards monopoly.

Scale economies are particularly critical for cloud computing. They offer clients access to state-of-the-art software, platforms and data infrastructure (storage, processing, etc.). Clients range from individual users (storing their photos and other files) to organizations, even governments, making use of information technology as a service. The provision of cloud services is highly concentrated at the world level, with four firms covering almost 70% of the market (Synergy Research Group, 2021). By the end of 2020, the four biggest providers were Amazon, Microsoft and Google from the US, and Alibaba from China. The three latter have steadily increased their market shares in recent years, while Amazon's share has stayed between 32 and 34%. This high degree of concentration reflects, besides network effects and first-mover advantage, huge investments in knowledge and infrastructure. These firms invest heavily both in research and development (R&D) and physical infrastructure, including data centres powered by supercomputers and connected by submarine cables.[3] What was initially an internal investment required to analyse their freely harvested data, resulted in a profitable business and a means to accelerate the adoption of digital technologies by every industry.

Public cloud computing market leaders are state-of-the-art software innovators. However, they are simultaneously platform leaders (Amazon and Alibaba in e-commerce, Google in search engines and social media while Microsoft owns GitHub and LinkedIn). By hosting other organizations on their platforms, tech giants can detect emerging new promising business areas and use the information strategically for further expansion of their own business.

1.2.3 National Security, Warfare and Artificial Intelligence

One direct reason for the technology war is that AI is regarded as a critical technology for warfare, including cyberwar, espionage and misinformation both in China and the US. This includes the rapid development of

---

[3] While new technologies such as edge computing where data are gathered and processed at the local level closer to clients are becoming more attractive in the era of AI, it is far from clear if it would weaken the dominance of the current providers. First, most edge computing applications require links to and interaction with centralized cloud computing. Second, the current big providers lead R&D in edge computing.

autonomous lethal weapon systems, where drones or submarines operate without human presence and, increasingly, with decisions taken through algorithmic processing of data. More generally, strategic considerations in relation to AI go in the same direction in China and the US.

In July 2017, China's State Council issued the New Generation Artificial Intelligence Development Plan (AIDP). The opening paragraph gives insight into how the Chinese leadership focuses on AI, both concerning economic development and international security:

> AI has become a new focus of international competition. AI is a strategic technology that will lead in the future; the world's major developed countries are taking the development of AI as a major strategy to enhance national competitiveness and protect national security. (Webster et al., 2017)

In the US, the National Security Commission on Artificial Intelligence (NSCAI) has as members several tech giants' top managers, including Eric Schmidt (former chief executive of Google's parent company, Alphabet) and Andy Jassy (current Amazon's CEO). It is a congressionally mandated, independent federal commission set up in 2019 "to consider the methods and means necessary to advance the development of artificial intelligence, machine learning and associated technologies by the United States to comprehensively address the national security and defense needs of the United States".[4]

1.2.4   Does China Constitute a Real Threat to US Supremacy?

The debate on China's new role as a challenger to the US is often focused on indicators showing the growing strength of China—one such observation is that aggregate GDP in PPP is now bigger in China than in the US. Another indication is that there were 36 Chinese corporations and only 30 US corporations among the 100 companies that gained the most in terms of market value since the pandemic began.[5] Yet, this exaggerates the economic power of China. The US remains far ahead in terms of economic and military strength. Moreover, while China has had a uniquely high rate of economic growth, the income gap is significant.

---

[4] https://www.nscai.gov/.

[5] https://www.ft.com/content/f8251e5f-10a7-4f7a-9047-b438e4d7f83a.

In US dollars, the GDP-per capita of China was around 1/6 of the US (around respectively 10,000 vs. 65,000 USD) in 2019.[6]

Also, when it comes to military strength, US dominance is obvious. According to EU statistics, the US allocates 610 billion USD, corresponding to 3.1% of its GDP, for military purposes. The corresponding figures for China are 228 billion USD and 1.9%. While the US has 20 aircraft carriers, China has 2. The US has 6450 nuclear warheads while China has 280. China only comes ahead of the US in terms of manpower—2.3 vs. 1.2 million active personnel—, but this is mostly a reflection of the size of their population. In China, 0.2% of the total population is engaged in military service while the share is 0.4% in the US.[7]

Although China remains far behind in terms of economic and military strength, it has been particularly ambitious when it comes to building a strong national innovation system, and in mobilizing investments in education, research and technology. The US investment in R&D increased 4.3% per year from 2000 to 2017. During the same period, China's R&D spending was growing by more than 17% per annum. R&D expenditure expanded threefold between 2000 and 2017 in China, rising from $722 billion to $2.2 trillion. In 2017, the US and China accounted for nearly half of the world's global R&D with 25 and 23%, respectively (National Science Foundation, 2020).

Therefore, everything points to an interpretation where the growing tension between the two countries, more than anything else, reflects a competition between core states and corporations shaping two big ambitious national innovation systems, both focused on winning the race in the strategic technologies of the ICT's second phase.[8]

[6] https://data.worldbank.org/indicator/NY.GDP.PCAP.CD?locations=CN.

[7] https://armedforces.eu/compare/country_USA_vs_China.

[8] Experts on geopolitics have warned that there is a risk for the US and China to run into the Thucydides Trap first identified by the ancient Greek historian Thucydides. As he explained, "It was the rise of Athens and the fear that this instilled in Sparta that made war inevitable". The past 500 years have seen 16 cases in which a rising power threatened to displace a ruling one. Twelve of these ended in war.

1.2.5   Competing National Innovation Systems

As we demonstrate in more detail in Chapters 7 and 8, in both the US and China, the state has been the guardian of their respective (data-driven and other) giant corporations.

In the case of the US, measures include colossal public investments in R&D and the strengthening and broadening of Intellectual Property Rights (IPRs). The authorization to patent living beings and software as wells as TRIPS agreement are two illustrative examples. They have been complemented with a weakening of antitrust policies leaving more room for harvesting monopoly rents from the home market (Glick, 2019; Kahn, 2017). The most recent example is the surge of protectionist trade measures and initiatives to restrict Chinese companies' access to advanced technologies.

The Chinese government has also used substantial resources to create a strong knowledge base for the inception of *made in China* intellectual monopolies. Big home markets attracting foreign firms have been used to negotiate access to foreign knowledge and to speed up technological learning from abroad. Since 2006, China follows an independent innovation strategy. China has combined supportive innovation policy with protectionist measures, such as reducing foreign access to the internet. China's "Great Fire Wall"—which blocks access for US companies to internet-based business—has favoured local corporations' development while impeding the access to the Chinese market for foreign firms, such as Amazon and Google (Azmeh et al., 2020; Mueller, 2011; Wu & Gereffi, 2018).

1.2.6   New Patterns of Globalization and Trade

The impact of the US–China conflict is global. Coupled with the relevance of AI and the other ICT second phase technologies, we are witnessing a new pattern of globalization.

Global value chains (GVC), where countries specialize in specific segments of a good or service supply chain, have been important for the growth of international trade at least since the beginning of the 1990s. However, since 2008, GVC trade expansion has stalled (Shin, 2019; World Bank, 2020, p. 19). Developments in China and the US explain much of the recent decline. Since 2010, Chinese companies have reduced the use of imported components in their exports. The US trade war and China's response have led to more abrupt disruptions of supply chains (Ciuriak, 2019; Ernst, 2020). More recently, the Covid-19 crisis contributed to the decline, and it also made national governments more aware of the risk of being dependent on supply from abroad.

In the same period, we have witnessed a different kind of globalization, characterized by rapid growth in trade of knowledge-intensive services and especially in digital services (WIPO, 2017; WTO, 2019). Since the financial crisis, world exports of manufactured goods grew by 2.3% per year while exports of business services grew by 3.9%, with ICT services growing 8% per year between 2008 and 2018. Markets for AI and cloud computing services are growing even more rapidly, and projections point to annual growth rates of respectively 37 and 15%.[9] In an era with stagnation in most industries, it is obvious that a strong technological and commercial base in digital services is critical for the growth potential of national economies.

UNCTAD's first report on the digital economy attempts to quantify the impact on economic growth of this new stage in the ICT revolution (UNCTAD, 2019). It finds that, while it is now possible to measure activities related to the manufacturing of ICT goods with some accuracy, this is not the case of the most rapidly expanding areas such as production and trade of digital services. Hence, the report makes use of alternative methods to assess impact, pointing to the explosive growth of market capitalization for the handful of giant companies that dominate digital services. The growing value of stocks reflects financial investors' expectations that digital giants will dominate their markets and therefore deliver greater profits.

Data on market capitalization confirm that the US and China form the geopolitical cores of digital capitalism, and that their dominance has increased since the financial crisis, both as home to giant corporations and in terms of aggregate market value. By March 31, 2019, seven of the world's top ten companies in market capitalization were tech giants, GAFAM, Alibaba and Tencent. They accumulated 23% of the market capitalization of the global top 100 companies in 2019. By mid-2020, the same 7 companies ranked in the top 10 and had all increased their market

---

[9] https://www.marketsandmarkets.com/Market-Reports/cloud-computing-market-234.html and https://www.globenewswire.com/news-release/2020/06/04/2043624/0/en/Global-Artificial-Intelligence-Market-Report-2020-to-2030-COVID-19-Growth-and-Change.html.

Table 1.1 Market capitalization (December 31, 2019) and liquid assets (annual reports 2019)

Market value world ranking	Market value Billion US Dollar	Liquidity Billion US Dollar
No 2. Apple	1305	205.9
No 3. Microsoft	1203	134
No 4. Alphabet (Google)	923	119.7
No 5. Amazon	916	55
No 6. Facebook	585	54.9
No 7. Alibaba	569	28.8
No 9. Tencent	461	29
Baidu	43.7	21.2
Huawei	n.a	13.06[a]

Source Authors' compilation based on Compustat and company reports
[a] https://www.huawei.com/en/annual-report/2019

capitalization. As we discuss in Chapter 6, recent downward movements in Chinese giants' market capitalization can mainly be explained by the conflict between the Chinese state and its tech giants.

1.3   PRESENTING THE CORPORATE PROTAGONISTS FROM THE US AND CHINA

GAFAM companies and their Chinese counterparts (BATH) are the corporate leaders of the digital economy. Until Saudi Aramco's IPO in December 2019, GAFAM were the biggest companies in the world in terms of market capitalization and, by 2021, Apple and Microsoft had surpassed Saudi Aramco. GAFAM have billions of liquidities (cash, cash equivalents and marketable securities) at their disposal. BATH constitute the only serious challenge for the US giants, but they still come behind in terms of market capitalization and a major difference is that, with the exception of Huawei and to a lesser extent Tencent, most of their customers are to be found within China (Table 1.1).[10]

---

[10] Since Huawei is not a public company—it is owned by its employees—there is no way to determine its market capitalization.

To get an idea of their giant scale, by 2019, the biggest company with head office in Europe, Nestlé, was ranked 13th worldwide in market value (USD 292 billion) (PWC, 2019).

The amount of liquidity is relevant since it represents financial strength, including the capacity to directly acquire other companies, to get quick access to new technologies and retain market dominance. To illustrate, Microsoft's liquidity (134 USD billion) can be compared to the market capitalization of Spotify (30 USD billion), Europe's biggest data services and tech company.

1.3.1   Mechanisms Behind World Dominance

There are different interpretations of the mechanisms that have led to the world dominance of GAFAM and BATH; we explore them in Chapters 2 and 3. Some explanations focus on their role as platform companies enjoying network effects and harvesting data. Other explanations point to their technological leadership and their massive investments in R&D. A third explanation has to do with the building of tangible infrastructure (not least in the form of data centres and cloud services and, in the case of Amazon, huge warehouses and distribution systems). A fourth refers to how they organize corporate innovation systems (networks of subordinate organizations). As we will demonstrate, none of these explanations can stand alone. A combination of factors underlies the dominance of GAFAM and BATH with prohibitive barriers to entry as result.

Another set of factors underlying the success of these corporations are the different forms of support that they get from their respective home states and their access to knowledge resources within and beyond their respective national innovation system. Both GAFAM and BATH benefit from easy access to public research (paid by taxpayers) as well as from direct state policies supporting their global dominance in strategic technologies. Beneficial tax structures and, as we mentioned earlier, lax antitrust policies in the US have underpinned GAFAM's emergence as intellectual monopolies.

1.3.2   Investments in Research and Development and Patenting

A substantial share of GAFAM's and BATH's income is invested in R&D and their economic power is primarily based on ownership and control of intangible assets. As reported in Table 1.2, they use significant shares

Table 1.2   R&D Expenditure and number of US patents

World ranking in terms of company spending on R&D	R&D Billion US Dollar	R&D Share of income %	Number of granted patents in the US in 2019
1. Amazon.com	22.6	12.7	1100
2. Alphabet (Google)	18.3	15.3	6602
4. Microsoft	14.7	13.4	7350
7. Apple	12.4	5.4	4362
12. Facebook	9.0	18.4	1265
6. Huawei	12.7	13.9	26,956
29. Alibaba	4.8	9.9	5010
54. Tencent	2.9	7.3	6337
74. Baidu	2.0	15.4	3815

Source Scoreboard (2019) and Google patents

of their annual income for investments in R&D and they are among the most active in terms of patenting worldwide.

It is noteworthy that Amazon's and Alphabet's (Google) R&D expenditures are higher than the total R&D expenditure in the Netherlands (15.9 USD Billion) and Sweden (17.1 USD Billion) while Microsoft comes close behind. Moreover, GAFAM and Huawei excel in applied and granted US patents. With the exception of Facebook, these companies belonged to the 20 companies granted most patents in the US in 2019.[11]

1.3.3   Pairwise Comparisons Between the Activities of US and Chinese Giants

At first glance, it seems as if there is a Chinese counterpart for each US tech giant. But while some activities overlap, a closer look points to important differences.

[11] https://www.uspto.gov/web/offices/ac/ido/oeip/taf/topo_19.htm.

1.3.3.1   *Amazon vs. Alibaba*

Their resemblance is apparent. They both began their operations by the end of last century (1994 Amazon and 1999 Alibaba) and they have become e-commerce behemoths with few serious competitors in their respective markets. Amazon stands for 39% of all US e-commerce sales and over 30% of the global market, and Alibaba for 58.2% of all retail e-commerce in China. Now, they compete in the cloud servicing market, each leading its respective country, although Amazon is the absolute global leader.

Even though their mode of operation looks similar at first glance, they are different in important respects. Amazon is not only a market-place but sells directly to consumers and operates its own warehouses, while Alibaba is only a middleman between buyers and sellers (from small businesses to brand manufacturers). Alibaba operates through several e-commerce sites aimed at different types of sellers. Alibaba's Taobao is one of its most profitable market platforms and is responsible for more than 80% of Alibaba's sales. Alibaba.com is its international online whole-sale marketplace and AliExpress is a global marketplace for international consumers. Another difference is the importance of Alibaba's e-payments and financial business, channelled through the Ant Group.

1.3.3.2   *Baidu vs. Google*

Searh engines are the original business of both companies and they both generate revenue primarily through online advertising and offering a wide set of web services. However, while Google concentrates most of the global market for internet search (90%), Baidu is dominant only in China, where it concentrated 79.9% of the online search queries, as of June 2021.[12] Google ranked a distant fourth in China's online search market, with a 2.4% share. To a significant degree, this global market distribution reflects China's firewall that blocked Google, but another factor has been Baidu's understanding of local Chinese language and culture. Baidu's search algorithms place much relevance to the context in which the words are used.

Another market where they compete is intelligent assistants. Here, Google is second after Amazon with almost a quarter of the market and

---

[12] https://www.statista.com/statistics/253340/market-share-of-search-engines-in-china-pageviews/.

Baidu follows with a 10% share.[13] Both companies are also investing heavily in autonomous driving, with Baidu explicitly backed by the Chinese government with the aim to make it a world leader.

Among their differences in business orientation, Google's Android is the dominant handset operating system, while Baidu does not compete in this market. Another difference is that while Google is strongly investing in digital healthcare, Alibaba and Tencent are the most active in this area in China.

1.3.3.3   Apple vs. Huawei

Apple and Huawei are in direct competition when it comes to the smartphone market, but they started in different lines of business, Apple entered the market selling state-of-the-art handsets and Huawei low-cost phones. However, as Huawei's technological capabilities improved, it eventually started producing phones that could compete with Apple's technology. In the smartphone market, Huawei has overtaken Apple in global market share—between 2018 and 2019, Apple's market share fell from 15.7 to 11.8% while Huawei's expanded from 11.8 to 18.9%.[14]

A difference between them is their manufacturing structure. In 2004, Apple completed its manufacture outsourcing, mostly to China (and since 2018 also to Vietnam and India echoing the US–China conflict). Apple is the paradigmatic example of GVC leader. Huawei, on the other hand, still assembles its handsets partly in-house. This contributes to explaining why, by December 2019, Huawei had over 194,000 employees worldwide, while Apple had only 137,000.

When the market for mobile phones stagnated, Apple expanded its digital services and wearables businesses. In the meantime, Huawei had already become the world's wireless telecommunications leader, a market it entered in the early 2000s. However, it was not until it mastered 5G that its leadership became clear.

[13] https://www.statista.com/statistics/792604/worldwide-smart-speaker-market-share/.

[14] https://www.idc.com/promo/smartphone-market-share/vendor.

1.3.3.4   Facebook vs. Tencent

Facebook and Tencent are both in control of social networks and their annual earnings are at around the same level. But, while Facebook operates globally with more than 2 billion users in more than 100 countries and concentrates 66% of the global market, Tencent's close to one billion users are mainly located in China. They also differ in terms of how they make money. A big proportion of Facebook's income, more than 90%, emanates from advertising. Tencent's revenue comes mainly from user activities, especially from gaming—its share of income from advertising is quite small. Tencent is the world's leading online gaming company. Another difference with Facebook concerns its successful online payments' business where it competes with Alibaba. Facebook is attempting to enter this market too, as evidenced by Libra, first launched as a new global payment system but later scaled down to a more closed system.

1.3.4   Technological Cooperation and Competition Between Tech Giants

Our pairwise counterposing of a US and a Chinese company with some overlapping products gives a first impression of the corporate players in the global innovation race. Nonetheless, as will become clear from the analyses in Chapters 3–5, it does not mean that GAFAM firms experience their respective Chinese counterpart as their main challenger.

GAFAM firms combine technological collaboration among themselves with technological and market competition, and the same is true for BATH. For instance, the main competitor of Facebook is Google when it comes to advertising. Moreover, as we have just mentioned, Alibaba is Tencent's main competitor in online payment. There is also important technological collaboration between some GAFAM and BATH companies. The GAFAM firm not presented in the pairwise comparisons—Microsoft—has, for instance, engaged in a close cooperation with Baidu to establish a self-driving car. Other participants in this technological race are Google and recently Amazon. Furthermore, Microsoft is Amazon's main competitor in cloud computing.

These companies are not monopolies in the sense of the standard textbook definition because they compete and challenge each other in almost every market where they operate. They are also sometimes challenged by other companies. As we develop in Chapter 2, their monopoly is of a different kind; primarily, they monopolize (access to) knowledge, and this is why we refer to them as intellectual monopolies.

The emergence of intellectual monopolies makes the geopolitical conflict and the technology war utterly complex. The tendency towards decoupling has important negative impacts on the technological advance and economic results both for GAFAM and BATH companies. Since these companies can influence politics in their respective home country, this kind of technological interdependence may thus serve to dampen the heat in the emerging cold war. Nevertheless, as we further develop in Chapters 6 and 7, they remain *citizens* of their home country and, as the struggle between states becomes more intense, they increasingly get mobilized for the interests of their home country.

1.4   THE STRUCTURE OF THE BOOK

Both GAFAM and BATH build market dominance through access to data, information and knowledge. In Chapter 2, we present them as examples of an increasingly important category of firms that we refer to as *intellectual monopolies*. Intellectual monopolies are characterized by continuously using innovation to keep competitors at bay. Beyond monopolizing specific markets, they monopolize fields of knowledge. We distinguish different degrees of intellectual monopoly and identify data-driven intellectual monopolies as belonging to the highest degree. They have monopolized a new method of invention: the application of deep learning and neural networks to process big data, producing digital intelligence that locates the most promising new combinations of the existing elements of knowledge. The chapter considers the threats for society of tech giants' consolidation as data-driven intellectual monopolies.

While GAFAM and BATH use substantial resources to pursue research and develop new technologies in-house, they do not invent and innovate alone. Besides acquiring promising companies, their innovation activities reflect that they organize and dominate global networks of innovators, including subordinate but formally independent firms and universities. In Chapter 3, we propose that it is useful to focus on how each intellectual monopoly builds, redesigns and leads a (global) *Corporate Innovation System (CIS)*. The chapter studies selected US and Chinese tech giants' corporate innovation systems through an analysis of their scientific publications' co-authoring organizations, their participation in the open-source software environment and start-ups' acquisitions as a means to access technology. It is shown that the formation and reconfiguration of global CIS transforms the geography of global knowledge production and how tech giants' global outreach to customers, suppliers and users is combined with highly localized interaction within and between their distinct CIS.

Chapter 4 analyses AI as a *technological innovation system* and shows the central role of selected tech giants in developing AI in general as well as in functional and field applications. We use data on tech giants' publications to analyse their crucial role in developing and shaping the system. They exploit positive externalities and dominate development and diffusion, market formation and resource mobilization. Through their corporate innovation systems and through acquisitions, they draw upon external entrepreneurial efforts. As a result, they get a prominent role in guiding search and the direction of new applications. Their main weak point relates to their legitimacy reflecting growing public resistance to their use and abuse of data. This is why tech giants make major efforts to become less dependent on access to big volumes of data.

Chapter 5 takes inspiration from the literature on technological paradigms and technological trajectories and looks into the emergence and dynamics of the AI technology innovation system. We explore the technological convergence between two tech giants with quite distinct origins using lexical analyses of these companies' patents and scientific publications. We find that both Amazon and Microsoft have zoomed in their R&D efforts on deep learning and neural networks as well as on functional AI applications. We also find evidence of increasing centrality of harvesting, storing and processing data. R&D on cloud computing infrastructure is another area where both companies overlap. Given their dominant role in the AI TIS, we argue that these companies' priorities are indicative of prevailing directions within AI technological trajectories.

Chapters 6 explores China's technological *catching-up* with the US and Europe in relation to AI and how it challenges US hegemony. It is inspired by Christopher Freeman's work on how radical technical change opens up for shifts in world leadership. Focusing on artificial intelligence as a core technology of the second phase of the ICT revolution, we overview the strengths and weaknesses of China's *national innovation system*—first in general and then with focus on AI. The catching-up process is analysed as a co-evolution of the country's national innovation system and the CIS of Alibaba and Tencent. We conclude that China's catching-up in AI has been impressive even if it remains second to the

US in most important areas. A key factor underlying China's unique success is Alibaba's and Tencent's privileged access to data from the enormous Chinese market. The other side of this is the expansion of state surveillance and censorship.

Chapter 7 focuses on the role of the state and its interaction with tech giants in the US–China race for AI global dominance. Since 2018, the US government has combined measures to promote AI with technology embargos aimed at protecting its lead in AI. Simultaneously, tech giants operate globally. The co-evolution between state policy and corporate strategy, characterized by elements of consonance and discord, is at the core of the current transformations of the global innovation system. We use the concepts *techno-globalism* and *techno-nationalism* to capture fundamental changes in state involvement and international relations. The US and Chinese states' techno-nationalisms overlayer the general tendency towards privatization of knowledge described in Chapter 2. What is taking place may be seen as a double enclosure process undermining the global knowledge commons and open science.

We end in Chapter 8 analysing global governance, introducing alternative scenarios concerning tech giants, the US and Chinese states, data governance and innovation. On this basis, we advance policy recommendations and calls for social activism aiming at a less polarized future where technology is guided towards collectively solving societal, ecological and health challenges. While it may be argued that microelectronics was a general-purpose technology, manufacturing was its primary field of application, whereas AI can be applied in all sectors and not least in those crucial for human welfare, such as education and health. What's more, it may revolutionize the innovation process itself. Hence, the current geopolitical distribution of corporate capabilities in AI is unsustainable. AI evokes both dystopian and utopian perspectives and its breakthrough should engage people all around the world to use their imagination to think about how to construct a fair and sustainable society for all, where global governance plays a stronger role and where knowledge is widely shared within and across nation states.

REFERENCES

Azmeh, S., Foster, C., & Echavarri, J. (2020). The international trade regime and the quest for free digital trade. *International Studies Review, 22*(3), 671–692.

Ciuriak, D. (2019, January 11). A trade war fuelled by technology. *Opinion*.

Ernst, D. (2020). *Competing in artificial intelligence chips: China's challenge amid technology war*. Centre for International Governance Innovation.

Glick, M. (2019). Antitrust and economic history: The historic failure of the Chicago school of antitrust. *The Antitrust Bulletin, 64*(3), 295–340.

Kahn, L. (2017). Amazon's antitrust paradox. *The Yale Law Journal, 126*(3). https://www.yalelawjournal.org/note/amazons-antitrust-paradox

Mueller, M. L. (2011). China and global Internet governance: A tiger by the tail. In *Access contested: Security, identity, and resistance in Asian cyberspace* (pp. 177–194).

National Science Foundation. (2020). *2020 The state of U.S. science & engineering*. National Science Foundation.

PWC. (2019). *Global top 100 companies by market capitalisation*. PWC.

Shin, H. S. (2019). *What is behind the recent slowdown?* BIS.

Synergy Research Group. (2021). *Cloud market ends 2020 on a high while Microsoft continues to gain ground on Amazon*. Synergy Research Group.

UNCTAD. (2019). *Digital economy report 2019: Value creation and capture— Implications for developing countries*. United Nations.

Webster, G., Creemers, R., Triolo, P., & Kania, E. (2017, August 1). Full translation: China's 'new generation artificial intelligence development plan' (2017). *DigiChina*.

WIPO. (2017). *World intellectual property report 2017: Intangible capital in global value chains*. World Intellectual Property Organization.

World Bank. (2020). *World development report 2020: Trading for development in the age of global value chains*. World Bank Publications.

World Intellectual Property Organization. (2019). *WIPO technology trends 2019: Artificial intelligence*. WIPO.

WTO. (2019). *World trade statistical review 2019*. World Trade Organization.

Wu, X., & Gereffi, G. (2018). Amazon and Alibaba: Internet governance, business models, and internationalization strategies. In *International business in the information and digital age* (pp. 327–356). Emerald Publishing Limited.

CHAPTER 2

Tech Giants as Intellectual Monopolies

Abstract In this chapter, we introduce and further develop *the intellectual monopoly* concept and argue that US and Chinese tech giants are paradigmatic examples. We distinguish different degrees of intellectual monopoly and identify data-driven intellectual monopolies as belonging to the highest degree. They have monopolized a new method of invention: the application of deep learning and neural networks to process big data, producing digital intelligence that locates the most promising new combinations of the existing elements of knowledge. The chapter considers the threats for society of tech giants' consolidation as data-driven intellectual monopolies.

Keywords Intellectual monopolies · Intangible assets · Innovation · Big data · Degrees of intellectual monopoly

2.1 INTRODUCTION

There is a growing literature on how capitalism is moving towards a new era characterized by economic concentration based on intangible assets (Auvray et al., 2021; Chen et al., 2017; Crouzet & Eberly, 2018, 2021; Haskel & Westlake, 2018; Rikap, 2021; Schwartz, 2020). Control of access to information and knowledge have become crucial for dominant

C. Rikap and B.-Å Lundvall, *The Digital Innovation Race*,
https://doi.org/10.1007/978-3-030-89443-6_2

market positions and for the expropriation of rents. To capture this trend, Pagano (2014) has coined the concept *intellectual monopoly capitalism.*

Although some intellectual monopolies concentrate great portions of certain markets, the reference to monopoly is aimed at emphasizing how corporations establish and sustain exclusive control and access to knowledge and information (an advantage that often leads to high degrees of market concentration). Furthermore, we define monopoly not as a state but with reference to a process where the incumbents keep on building new barriers more rapidly than newcomers can overcome to break through the old ones.

US and Chinese tech giants are among world leaders in innovation, and they have dominant positions in domestic and world markets. We characterize them as intellectual monopolies whose control of knowledge goes beyond specific technologies and their corporate power goes beyond specific markets. Seven of the nine corporations studied (we exclude here Apple and Huawei) are prototype examples of intellectual monopolies because their core activity is to process data, information and knowledge.

Intellectual monopoly capitalism is the outcome of privatization, commodification and assetization of knowledge with company ownership of digital platforms contributing to the concentration of knowledge assets. In what follows, we use basic theoretical insights and concepts from the economics of knowledge and political economy to explain the ongoing transformation and assess its socio-economic consequences.

Section 2.2 gives empirical background and Sect. 2.3 presents mechanisms explaining the emergence of intellectual monopoly capitalism. Section 2.4 introduces a taxonomy distinguishing between different degrees of intellectual monopoly. As conclusion, we open up the discussion of who suffer the negative consequences of intellectual monopoly capitalism.

2.2   THE EMPIRICAL EVIDENCE

The trend towards increased corporate power based on intellectual capital is worldwide (IMF 2019—World Economic Outlook). It is, however, most pronounced within and most thoroughly documented for the US.[1]

[1] An exception is a recent OECD report that studies 26 countries and shows that markups are concentrated at the top of the markup distribution and that they are higher for digital-intensive sectors (Calligaris et al., 2018).

The literature indicates higher concentration at three interrelated levels. First, the overall share of value going to corporate profits is growing while the share of wages is declining. Second, there is concentration at the organization level, with a few firms concentrating a growing share of revenues and profits. Third, the companies driving this concentration are those intensive in the use of intellectual/intangible capital (Covarrubias et al., 2020; Haskel & Westlake, 2018; Orhangazi, 2018).

The share of US GDP going to corporate profits has risen significantly. Between 1985 and 2016, it rose from around 7.5 to over 11% (Shapiro, 2018). Second, there has been a rise in another mainstream indicator of market dominance, the ratio between price and marginal cost over the last decades. For instance, Hall (2018) finds that the weighted average ratio of price to marginal cost increased from 1.12 in 1988 to 1.39 in 2015. Third, various measures of broad industry concentration in the US economy have increased (Autor et al., 2017).

Corporate power is increasingly based on intangible capital (Haskel & Westlake, 2018). At the global level, it can be illustrated by the shift in market capitalization rankings. A dominant trend is the move from oil and gas companies (like Exxon Mobil and PetroChina) and banks (Bank of America, City Group and ICBC) to big-tech dominance. As we mentioned in Chapter 1, 7 of the top 10 corporations in market capitalization are tech giants (Google, Amazon, Facebook, Apple, Microsoft, Alibaba and Tencent).

The market capitalization ranking is increasingly dominated by the world's most active companies in terms of R&D investment. In 2009, only two of the top ten corporations in market capitalization were also among the top ten companies in terms of Business Expenditure on Research and Development (BERD). By 2019, this was true for 5 out of the top 10 corporations (European Commission, 2019; PWC, 2020).

Global innovation efforts are becoming concentrated in a small number of big companies. Since 2011, Clarivate has presented a ranking of the 100 most innovative companies worldwide. This strong core of organizations is dominated by multinationals (97 of the 100 organizations in 2020), mostly headquartered in the US. Apple and Microsoft have been ranked every year and, in 2020, the ranking included GAFAM, Huawei and Tencent.

The concentration of innovation efforts in a small number of global companies goes hand in hand with a concentration of BERD. By 2018, the ten companies with the highest BERD accumulated 13.5% and the top

100 concentrated 48% of world BERD (European Commission, 2018). That same year, the top 10 companies invested more than twice as much as the top 10 companies did in 2008.

2.3   Intellectual Property Rights and Assetization of Knowledge

Patenting and copyright have existed for centuries. However, it was not until the end of the twentieth century and, in particular, the twenty-first century that an exponential concentration of IPRs took place (see Fig. 2.1). While it has been seen as a method to stimulate innovation, its main impact on the economy is massive privatization and assetization of knowledge. This trend, where knowledge is transformed into assets, has triggered growth in intellectual rents.

The expansion in scope and space of IPRs paved the way for the rise of intellectual monopoly capitalism, a story that owns much to the US state (Rikap, 2021, Chapters 3 and 4). But in order to understand the roots

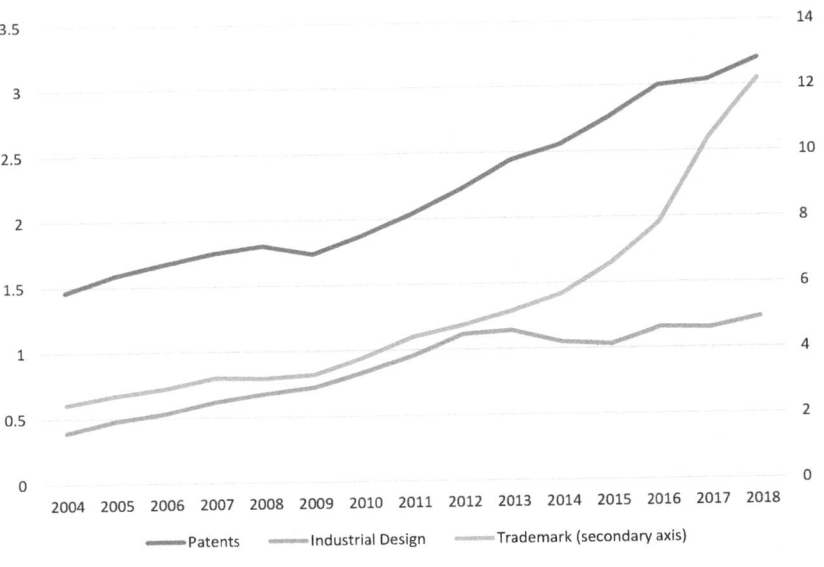

Fig. 2.1   IPRs application class counts for the top 20 offices (in millions) (*Source* WIPO)

of the corporate power of tech giants and other intellectual monopolies, it is necessary to focus on how they combine legal protection with other mechanisms such as secrecy and platform ownership.

2.4   Mechanisms Behind the Emergence of Intellectual Monopoly Capitalism

How did the world arrive at such extreme levels of market and knowledge concentration? To answer these questions, we proceed in two analytical steps. First, we analyse the surge in IPRs and how it has led to massive assetization of knowledge. Second, we show that for the giant firms other means of concentrating and controlling knowledge, such as secrecy and platform ownership, are at least as important.

2.4.1   *The Concentration of IPRs*

Corporate lawyers have been at the forefront fighting for extending and guarding IPRs (Bessy, 2019). This is not surprising since they are hired to ensure that the company remains as protected from competition as possible. To defend their actions, they use arguments developed by mainstream economists saying that it is costly to develop new ideas while it is almost costless to copy them leaving the innovator without compensation and without incentives to engage in innovation. To strengthen their case, they might add that without IPRs, companies would keep their new technology secret and thereby technological progress would be slowed down.

Innovation scholars who have studied the empirical reality of innovation processes, such as Boldrin and Levine (2008), disagree and characterize IPRs as an unnecessary evil because:

1. Most important innovations have taken place without the use of IPRs. And there are other and better ways to compensate innovators.
2. Only in exceptional cases is imitation costless—and in those cases the incentives to imitate would be weak. Normally, it takes major efforts to absorb and use new technology.
3. IPRs only compensate the winner of the innovation race. They neglect that innovations are combinations of existing elements of

knowledge and involve interaction between individuals and organizations. This is one reason why a system allocating all the gains to the firm that happens to win a patent race is neither fair nor efficient.

The use and abuse of legally guaranteed property rights are at the core of the emergence of intellectual monopoly capitalism (Boldrin & Levine, 2008; Pagano, 2014). Pagano (2014) makes an important distinction between private ownership of tangibles and knowledge ownership. The exclusion of access to and use of a tangible good is local and specific. This stands in contrast to the exclusion from using an element of knowledge, which is a global exclusion. He also draws parallels between the privatization of knowledge and the historical Enclosure of the Commons. Again, there is a major difference between the enclosure process, which only affected local communities, and the privatization of knowledge, which has global consequences such as increasing inequality between countries. Pagano (2014) also points out that while it has been argued that the enclosure could be seen as overcoming "the tragedy of the commons" (free access led to overexploitation of land), there is no such tragedy in the knowledge commons. Actually, in this case, the tragedy is that the restriction on the use of existing knowledge reduces the potential of learning (thus the production of new knowledge) from its use.

Overall, Pagano (2014) concludes that the accumulation of intellectual capital blocks innovation outside intellectual monopolies and that this is a major explanation of the prolonged economic crisis.

In any case, the use of IPRs is rising, and their rationale extends beyond protecting an innovation, including expansion in international markets and enforcing standards. Patenting is a widely spread practice to threaten and bargain with rivals—particularly frequent among the US tech giants to block potential competitors. Noel and Schankerman (2013, p. 514) found that rivals' patenting led to a reduction of firms' R&D investment, patenting and market value. The use of legal protection of knowledge has become a more important dimension of business strategies. The rates of patent applications and of charges for the use of intellectual property[2] grow more than the average global GDP (see Fig. 2.2). Furthermore, patent portfolios become increasingly concentrated in a small number of

[2] This indicator only sets a floor for legal innovation rents defined as the rents garnered from IPRs. It only considers payments between countries and does not include rents garnered from the sale of products that use patented technology.

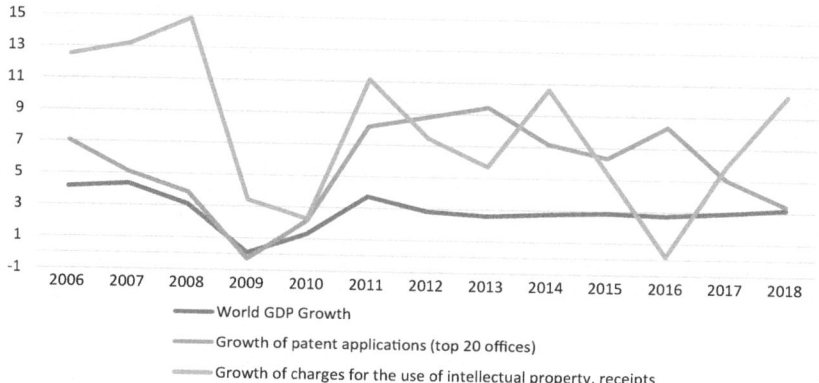

Fig. 2.2 World GDP, patent application and licences growth[3] (*Source* World Bank and WIPO)

giant firms typically rooted in high-income countries (Clarivate Analytics, 2020; Dutta et al., 2019), reflecting smaller firms' difficulties in absorbing costs of filing and defending patents (Munari, 2013).

A particular patenting strategy has been to fragment innovations in as many different patents as possible (called patent thickets). This strategy increases enforcement costs and chances to threaten those trying to copy (Ernst, 2016; Noel & Schankerman, 2013) and intellectual monopolies are large-scale patenting corporations, with the skills to face claims in courts and strong deterrence capacity. This combination contributes to high entry barriers for newcomers while favouring big players (Pagano, 2014). Giant corporations' concentration of resources, including legal expertise and sufficient funds to outsource this service in specialized consulting firms, gives them the upper hand in disputes on IPRs. As a result, capital accumulation increasingly becomes concentrated in intellectual monopolies.

---

[3] Annual data is presented as an average of year t and $t-1$ to smooth the cycle and get a better estimate of the trend.

2.4.2   Intellectual Monopolies Go Far Beyond IPRs

However, IPRs are only one—and not necessarily the most important—mechanism to prevent others from accessing knowledge. Summarizing multiple studies on the effects of patents on innovation incentives, Sampat (2018) concludes that these effects are highly sector-specific and that, except for pharmaceutical and chemical industries, patenting is not the most important appropriability mechanism to assure intellectual rents. The same results are achieved by Comino et al. (2019), specifically for ICT even if IPRs are also mounting in these fields.[4] Besides IPRs, according to Munari's (2013) literature review, protective mechanisms include secrecy, tacit knowledge, lead-time advantage and sale of complementary products, among others. The chosen mechanism depends on the nature of the innovation, firms and industry characteristics as well as institutional contexts. Firms use different mechanisms in complementary ways and tend to consider lead-time advantage and secrecy as the most effective mechanisms.

Regardless of the used mechanism, the private appropriation of intangibles such as information generates, in Veblen's (1908) words, artificial scarcity based on a right to exclude. The privatization of knowledge enable its owner to capture value in the form of rent reflecting monopoly power (Birch, 2019; Durand & Milberg, 2020; Foley, 2013; Kaplinsky, 1998; Pagano, 2014; Teixeira & Rotta, 2012).[5] Rentiership requires active behaviour towards property rights both in terms of the management of private assets and in terms of enforcing those rights (Birch, 2017).

Intellectual monopolies rely on IPRs when necessary, but they are not based exclusively upon legal ownership to knowledge. Our analysis of selected US and Chinese tech giants in Chapters 3–5 demonstrates that

[4] Considering patents quantity and quality, Clarivate Analytics top 100 global innovators include 38 hardware and electronics companies, 8 software companies and 8 telecommunications companies. Altogether, these figures represent an increase of 8 companies from 2018 (Clarivate Analytics, 2020).

[5] Our definition of rent builds on the one developed by the classical political economists for explaining ground rent (Ricardo and Marx). Rents differ from profits because the former is an appropriation of value that emerges from the monopolization of an (sometimes artificially) scarce resource that has been privately appropriated, whereas profits emerge from a production process that creates new value. Profits are new value created by workers in the production process kept by capitalists due to their ownership of the means of production, while rents represent a redistribution of value.

their market dominance and knowledge monopolies are based upon a combination of strategies.

Much new technological knowledge (and data) remains *secret* and there have even been arguments indicating that some of the most valuable knowledge is kept in this form. Dolata (2017) has pointed out that major innovations behind GAFAM companies tend to be kept behind closed doors. This is the case of Amazon's and Google's search algorithms (beyond the initial one owned by Stanford University even if its patent has already expired), Microsoft software packages and Facebook's social graphs.

In fact, US and Chinese tech giants combine legal protection and secrecy with extracting benefits from open-source software and from opening their platforms to freelance developers, as we further develop in Chapter 3. And, while iOS is not open-source, Apple also works as a gatekeeper deciding which apps will be offered for its iPhones, at what price and the fees it will charge to developers. The latter was the US Congress main claim against Apple during the recent Big Tech's investigation (US Congress. Subcommittee on antitrust, commercial and administrative law, 2020).

2.4.3 *Data-Driven Intellectual Monopolies Are Reinforced by Platform Ownership*

Platforms allow their owners to harvest data. They are widely used by what we will define as data-driven intellectual monopolies, from Amazon and Alibaba marketplaces to Tencent and Facebook social networks and Google's search engine. All these are examples of platforms where tech giants harvest constant streams of diverse data. Platforms enjoy network effects that contribute to expanding harvested data (Srnicek, 2017). Other sources of data do not necessarily rely on platforms. Online services, such as Gmail, Microsoft 365 and even IoT data may be harvested without a platform as mediator.

Conversely, not all platform companies have succeeded to establish themselves as intellectual monopolies. This can be illustrated by Uber. Uber introduced a new way to organize ride-hailing. It was a onetime breakthrough that was quickly copied by others (Lyft, Cabify, Bolt, etc.). This example points to the difference between the temporary advantage of one-time innovators with or without legal protection—what we call an intellectual monopoly degree 1—and higher degrees of intellectual monopoly. Furthermore, Uber depends on Google's cartographic technologies to operate and its data sources are not as diverse and multipurpose as those of tech giants. In this book, we focus on tech giants, which are intellectual monopolies of a higher degree. They use their position as platform owners to innovate further and to produce digital intelligence (see Chapters 4 and 5).

2.5   DEGREES OF INTELLECTUAL MONOPOLY—A TAXONOMY

Intellectual monopoly capitalism may be defined as the epoch where capital accumulation is increasingly driven by the systematic concentration of knowledge in a small number of corporations. Intellectual monopolies emerge from the capacity of certain firms to monopolize innovation within a branch or industry (Durand & Milberg, 2020; Pagano, 2014; Rikap, 2018; Schwartz, 2016). While Pagano (2014) ascribes this to the concentration of IPRs within these firms, we have emphasized how this is combined with other mechanisms and with the centralization of (processed) data and digital intelligence.

In the previous section, we identified different mechanisms that enable companies to monopolize knowledge. IPRs play a major role, but we also recognized that secrecy—including the accumulation of experience-based tacit knowledge—is another knowledge monopolization mechanism. Intellectual monopolies combine these mechanisms favouring one over the other according to the specificities of each production and innovation process.

Below we argue that the scope and consequences of monopolizing knowledge will differ according to the degree of generality of monopolized knowledge. When intellectual monopoly is based exclusively on IPRs, it is temporary and technology-specific and we refer to it as intellectual monopoly of first degree (IM1). When it is combined with other mechanisms leading to an enduring dominance in a specific technology, we refer to it as IM2. When a company establishes technological dominance in relation to a *general-purpose technology* (GPT) we refer to it as IM3 while the most extreme form of intellectual monopoly (IM4) is one where a company or a handful of companies dominate the development of a *GPT that is at the core of the innovation process itself, thus monopolizing an innovation process.* Most of the theoretical literature on competition and antitrust principles have given attention to consequences of IM1 (see for instance Boldrin & Levine, 2008; Lambert, 2019; Pagano, 2014). In the rest of this section, we further elaborate on the three other degrees of intellectual monopoly.

2.5.1   Intellectual Monopoly of Second Degree

A corporation that monopolizes knowledge within a specific technology will garner rents from it. IM2 are companies engaged in permanent innovation, arriving systematically first, building on previous achievements, expanding the gap with others in that field.

Within AI, State Grid Corporation of China and Siemens are among the top 20 organizations in terms of patent portfolios. However, each of them builds its intellectual monopoly in relation to sector-specific technologies. The former grew its patent filing by 70% annually between 2013 and 2017, but with AI applications strongly focused on energy, which makes sense since it is the world's biggest utility company. Siemens is the leader in AI patents applied to life and medical sciences, in particular related to medical images. Other examples within AI are Toyota and Bosch, concentrating AI patents on transportation applications (World Intellectual Property Organization, 2019).

2.5.2   Intellectual Monopoly of Third Degree: The GPT Intellectual Monopoly

While IM2s monopolize knowledge and act as intellectual rentiers, the scope of their monopoly is limited. This is not the case for IM3. These are corporations that have monopolized a GPT. As defined by Antonelli (2014, p. 23), GPTs emerge from the complementarity and interdependence of multiple innovations. They can be applied to numerous and diverse production processes, and they have the potential to permeate and shape manifold specific technologies.

The greater scope of the intellectual monopoly achieved by a corporation monopolizing a GPT extends the fields for harvesting intellectual rents. Furthermore, the effects that this degree of monopoly have on knowledge as commons are way beyond those of first- and second-degree intellectual monopolies because of the widening of potential applications that will be curtailed and the expansion of those that will have to pay rents to the intellectual monopoly in order to access essential knowledge.

Concerning the first ICT phase, which was itself a GPT, Microsoft's Windows is an example of a long-term and well-established third-degree intellectual monopoly. A dominant position that not even China has managed to challenge.

Smartphones are another example of a GPT monopolized by three high-tech giants that are also the undisputed leaders of the smartphones' market: Apple, Huawei and Samsung (WIPO, 2017). WIPO (2017) shows that smartphones' related patents have grown steadily since the end of the last century and that there is high correlation between value capture and the use of intellectual property in this industry. According to this report, "Apple keeps 42 percent of the retail price of each iPhone sold (or USD 270), Huawei 42 percent (USD 203) and Samsung 33 per cent (USD 221.76)" (WIPO, 2017, p. 100).[6]

In ICT second phase, the fear of the US—and in fact of all the Western world—in relation to Huawei's 5G supremacy illustrates the geopolitical importance of third-degree intellectual monopolies. Other cases that we explore in Chapters 4 and 5 are cloud servicing, currently in few hands within tech giants, and quantum computing on its way to become a third-degree intellectual monopoly in the hands of Microsoft and Google.

2.5.3   Intellectual Monopoly of Fourth Degree: The Data-Driven Intellectual Monopoly

Given its multiple applications, AI has been considered a GPT (Brynjolfsson et al., 2017; Dosi & Virgillito, 2019). Yet, within this GPT, as we explained in Chapter 1, machine learning approached as a combination of deep learning and neural networks[7] is indispensable to transform big data into economic outcomes. This is the process that UNCTAD

---

[6] Concerning standard-essential patents (SEP), Samsung (13.5%) and Huawei (9.9%) have the highest share of 4G related patents. Surprisingly, Apple only holds 1.7% of 4G SEP. However, as the report keeps explaining, secrecy and other less formal appropriation mechanisms play an important role.

[7] These are the fastest-growing AI technologies, as evidenced by patent application growth (World Intellectual Property Organization, 2019).

(2019) labelled "digital intelligence" and its specificity lies in how algorithms learn by analysing big data. Therefore, it is not surprising that the major owners of digital platforms are also lead innovators when it comes to developing AI (see Chapters 4 and 5).

Unlike previous GPT, digital intelligence has the potential to become *a method of invention* (Cockburn et al., 2018). Deep learning and neural networks can speed up the process of innovation using algorithms to locate the most promising new combinations of the existing elements of knowledge. AI, in particular machine learning, is changing the innovation process itself, with digital intelligence offering almost unlimited applications.

If this general-purpose method of invention is concentrated in a few hands, the potential for further innovating will be privatized with barriers for newcomer enterprises. Once this happens, the degree of centralization of innovative power may expand exponentially. In Chapter 4 we analyse, how Google, Amazon, Microsoft, Alibaba and Tencent are concentrating efforts to establish themselves as world leaders in machine learning as a method of invention. With only a handful of tech giants dominating the largest and most diverse datasets as well as innovations in deep learning and neural networks to analyse that big data, it can be expected that the tendency towards concentration of innovation in a small number of companies will become even more extreme in the future.

2.5.4 On the Need to Rethink Innovation, Monopoly and Antitrust

Innovation scholars such as Bronwyn Hall, following the tracks of Penrose (1951) and Machlup (1958), remain sceptical to the rationale behind intellectual property rights. They are, however, reluctant to call for a general abolition because there is no conclusive argument neither in favour nor against the "belief that the patent system has promoted the progress of the technical arts and the productivity of the economy". Furthermore, since it was already introduced, it would therefore be irresponsible to abolish it as much as it would be irresponsible to introduce it if it had not already been introduced (Machlup, 1958, pp. 79–80). Innovation studies scholars' criticism is tamed by their inspiration from Joseph A. Schumpeter who argued, in his late work, that big firms with market power are more active and successful in organizing innovation than small firms engaged in intense price competition (Schumpeter, 1942).

In a world with a handful of global companies moving from IM3 towards IM4 there is a need for a more radical criticism of intellectual monopoly. IPRs allow the right-holder to collect legal intellectual rents and, as the system is generalized, it shrinks the space for competitors' innovative activities and slows down the spread of new ideas. This is at the roots of Pagano's (2014) argument against intellectual monopoly. We add that the generalized transformation of knowledge into an intangible asset undermines the institutions of the learning economy.

Innovation is, by definition, an uncertain process—if the outcome of the process was known in advance it would not be an innovation. To overcome technological uncertainty, the innovator links up with suppliers and knowledge institution and, to overcome market uncertainty, it establishes long-term relationships with customers and users (Christensen & Lundvall, 2004). Extending private ownership to knowledge transforms these relationships. Untraded interdependencies become traded, formalized and turned into formal (asymmetric) contracts. Trust relationships are undermined and, as result, transaction costs escalate and the space for innovation based on learning by doing, using and interacting is reduced.

In the 1990s, the debate on the Solow-paradox, pointed to the fact that the rate of productivity growth remained modest while information technology was spreading rapidly. Today a related but somewhat broader debate is taking place regarding "secular stagnation". Some scholars explain the slowdown in productivity growth from the demand side (Summers, 2014, 2016) while others refer to the supply side (Gordon, 2015). With the exception of Gordon (2015), participants assume that the stagnation takes place in spite of rich technological opportunities and a high rate of innovation. We would argue that this reflects a superficial view of the innovation process that neglects unintended consequences of assetization and concentration of knowledge. Among the few recent exceptions, Durand and Gueuder (2018) argue that intellectual monopolies fuel weak investment since the monopolization of knowledge by and beyond IPRs blocks potentially profitable opportunities. Hence, at the global level, while intangible accumulation is associated with market power that increases prices per unit of investment, it does not translate to the overall growth of the economy (Rabinovich, 2020).

2.6   Conclusion: Corporate Winners and Losers

The concentration process presented in this chapter is reflected in technological differentiation between firms. The formation of intellectual monopoly capitalism is a dual process because the technological enhancement of the leaders simultaneously reduces the remaining firms' possibilities and incentives to invest in future innovations (Levín, 1997; Pagano, 2014). While current monopoly positions are not given once and for all, the barriers hindering competitors from entering have become extremely high and the incumbents use intellectual rents to invest in new (or reinforce existing) barriers. At the system level, this results in a slowdown of economic growth. However, at the level of the intellectual monopoly corporation, intellectual rents keep expanding. Innovations, taking place in-house but also at the level of intellectual monopoly's corporate innovation systems (see Chapter 3), thrive.

In this chapter, we have further developed the concept *intellectual monopoly capital*. We have shown how corporate winners monopolize knowledge by concentrating IPRs but also by using other protection mechanisms. We have also proposed that they establish themselves as intellectual monopolies of different degrees. They range from companies monopolizing single technologies, to companies that have monopolized a GPT with the particular case of data-driven intellectual monopolies, which have monopolized a GPT that is also a new method of invention. The latter has the potential to expand knowledge monopolization to multiple and the most diverse technologies (see Chapter 4). We classify US and Chinese tech giants as 3.0 intellectual monopolies, with at least some of them aiming at becoming fourth-degree intellectual monopolies.

Flipping the coin, losers have multiplied. The reign of intellectual monopolies (mostly rooted in the US but operating in global markets) has detrimental effects on innovation at the global level, increases world inequalities and leads to stagnation. It has blocked innovation for local firms in key technologies, with the peculiar exception of China where policies were put in place to impede or limit US data-driven intellectual monopolies access. In turn, as we will argue in Chapter 6, these policies have been crucial for the establishment of Chinese intellectual monopolies.

The production of tacit and embedded knowledge reflects processes of learning by doing, using and interacting. In an economy without innovation (or with innovation enclosed by intellectual monopolies), this would, by itself, give an advantage to incumbents. With ubiquitous technological change, entrepreneurship and innovation, such advantages would not last for long, however. Incumbents would get stuck in routines and innovation would open windows of opportunity for newcomers. But these mechanisms of capitalist churning may become blocked when incumbents are tech giants engaged in relentless innovation and with dominance in developing key technologies on the way to become fourth-degree intellectual monopolies.

Furthermore, the emergence and spread of data-driven intellectual monopolies have a major impact on the industrial structure worldwide. Amazon's e-commerce undermines big and small retailing, Google and Facebook's growing roles in advertising have a negative impact not only on advertisement agents but on all kinds of media, including newspapers.

Moreover, it may be argued that the expansion of platform companies leads to more intensive competition in the exposed areas and that this is to the advantage of consumers. However, the other side of the transformation takes place in the labour market, where the main tendency is to destroy skilled and stable jobs and create casualized jobs, often with fewer skill requirements. There is growing evidence that this mechanism is one important explanation of growing income inequality.

All in all, there is little doubt that the concentration of intellectual property in high-income countries and giant firms has made the world economy more unequal and polarized. Indeed, Pagano (2014) sees intellectual monopoly capitalism as a major explanation of growing inequality within and across countries. To this, we add the fundamental democratic problem that, as fourth-degree intellectual monopolies consolidate, the direction of technologies shaping our society in fundamental dimensions (including health, education and ecological matters) will be determined by a handful of global corporations.

The next chapter will elaborate on how intellectual monopolies build global *corporate innovation systems* subordinating business and academic organizations worldwide while still relying on their national innovation systems. Corporate winners do not innovate alone. They draw upon information from clients and customers worldwide and they benefit from knowledge production and innovation activities outside the firm, taking place at universities and other academic research institutions, including knowledge-intensive ventures (typically ICT start-ups in high-tech or biotechnology start-ups in pharma).

REFERENCES

Antonelli, C. (2014). *The economics of innovation, new technologies and structural change*. Routledge.

Autor, D., Dorn, D., Katz, L. F., Patterson, C., & Van Reenen, J. (2017). *The fall of the labor share and the rise of superstar firms*. Centre for Economic Performance, LSE.

Auvray, T., Durand, C., Rabinovich, J., & Rikap, C. (2021). Corporate financialization's conservation and transformation: From Mark I to Mark II. *Review of Evoluationary Political Economy*. https://doi.org/10.1007/s43253-021-000 45-4

Bessy, C. (2019). *The transformations of conventions for patent use and the role of legal intermediaries*.

Birch, K. (2017). Financing technoscience: Finance, assetization and rentiership. In D. Tyfield, R. Lave, S. Randalls, & C. Thorpe (Eds.), *The routledge handbook of the political economy of science* (pp. 169–181). London, UK: Routledge.

Birch, K. (2019). Technoscience rent: Toward a theory of rentiership for technoscientific capitalism. *Science, Technology, & Human Values, 45*(1), 3–33.

Boldrin, M., & Levine, D. K. (2008). *Against intellectual monopoly*. Cambridge University Press.

Brynjolfsson, E., Rock, D., & Syverson, C. (2017). *Artificial intelligence and the modern productivity paradox: A clash of expectations and statistics*. National Bureau of Economic Research.

Calligaris, S., Criscuolo, C., & Marcolin, L. (2018). *Mark-ups in the digital era* (OECD Science, Technology and Industry Working Papers). OECD.

Chen, W., Gouma, R., Los, B., & Timmer, M. P. (2017). *Measuring the income to intangibles in goods production: A global value chain approach*. World Intellectual Property Organization-Economics and Statistics Division.

Christensen, J. L., & Lundvall, B. Å. (2004). *Product innovation, interactive learning and economic performance*. Elsevier Jai.

Clarivate Analytics. (2020). *Derwent top 100 global innovators 2020*. Clarivate Analytics.

Cockburn, I. M., Henderson, R., & Stern, S. (2018). *The impact of artificial intelligence on innovation*. National Bureau of Economic Research.

Comino, S., Manenti, F. M., & Thumm, N. (2019). The role of patents in information and communication technologies: A survey of the literature. *Journal of Economic Surveys, 33*(2), 404–430.

Covarrubias, M., Gutiérrez, G., & Philippon, T. (2020). From good to bad concentration? US industries over the past 30 years. *NBER Macroeconomics Annual, 34*(1), 1–46.

Crouzet, N., & Eberly, J. (2018). Intangibles, investment, and efficiency. *AEA Papers and Proceedings, 108*, 426–431.

Crouzet, N., & Eberly, J. C. (2021). *Intangibles, markups, and the measurement of productivity growth* (NBER WORKING PAPER SERIES).

Dolata, U. (2017). *Apple, Amazon, Google, Facebook, Microsoft: Market concentration-competition-innovation strategies* (SOI Discussion Paper). Stuttgarter Beiträge zur Organisations-und Innovationsforschung.

Dosi, G., & Virgillito, M. E. (2019). Whither the evolution of the contemporary social fabric? New technologies and old socio-economic trends. *International Labour Review, 158*(4), 593–625.

Durand, C., & Gueuder, M. (2018). The profit-investment nexus in an era of financialisation, globalisation and monopolisation: A profit-centred perspective. *Review of Political Economy, 30*(2), 126–153.

Durand, C., & Milberg, W. (2020). Intellectual monopoly in global value chains. *Review of International Political Economy, 27*(2), 404–429.

Dutta, S., Lanvin, B., & Wunsch-Vincent, S. (2019). *The global innovation index 2019: Creating healthy lives—The future of medical innovation.* World Intellectual Property Organization (WIPO), Geneva, Switzerland.

Ernst, D. (2016). China's bold strategy for semiconductors–zero-sum game or catalyst for cooperation? *East-West Center Working Papers: Innovation and Economic Growth Series.*

European Commission. (2018). *EU R&D SCOREBOARD.* European Commission.

European Commission. (2019). *EU R&D SCOREBOARD.* European Commission.

Foley, D. K. (2013). Rethinking financial capitalism and the "information" economy. *Review of Radical Political Economics, 45*(3), 257–268.

Gordon, R. J. (2015). Secular stagnation: A supply-side view. *American Economic Review, 105*(5), 54–59.

Hall, R. E. (2018). *New evidence on the markup of prices over marginal costs and the role of mega-firms in the us economy.* National Bureau of Economic Research.

Haskel, J., & Westlake, S. (2018). *Capitalism without capital: The rise of the intangible economy.* Princeton University Press.

Kaplinsky, R. (1998). *Globalisation, industrialisation and sustainable growth: The pursuit of the nth rent.* Institute of Development Studies, University of Sussex.

Lambert, T. E. (2019). Monopoly capital and innovation: An exploratory assessment of R&D effectiveness. *International Review of Applied Economics,* 1–14.

Levín, P. (1997). *El capital tecnológico.* Catálogos. http://www.econ.uba.ar/www/institutos/economia/Ceplad/elcaptec.htm

Machlup, F. (1958). *An economic review of the patent system.* US Government Printing Office.

Munari, F. (2013). Review of literature on the use and impact of IPRs at the firm level: Patents, trademarks and designs. In *EUROPEAN OBSERVATORY FOR THE INFRINGEMENT OF INTELLECTUAL PROPERTY RIGHTS (ED.)*. https://Oami.Europa.Eu/Ohimportal/de/Web/Observatory/Observatory-Publications

Noel, M., & Schankerman, M. (2013). Strategic patenting and software innovation. *The Journal of Industrial Economics, 61*(3), 481–520.

Orhangazi, Ö. (2018). The role of intangible assets in explaining the investment–profit puzzle. *Cambridge Journal of Economics, 43*(5), 1251–1286.

Pagano, U. (2014). The crisis of intellectual monopoly capitalism. *Cambridge Journal of Economics, 38*(6), 1409–1429.

Penrose, E. T. (1951). *The economics of the international patent system*. Johns Hopkins Press.

PWC. (2020). *Global top 100 companies by market capitalisation*. PWC.

Rabinovich, J. (2020). Financialisation and the 'supply-side' face of the investment-profit puzzle. *Journal of Post Keynesian Economics*, 1–29.

Rikap, C. (2018). Innovation as economic power in global value chains. *Revue D'économie Industrielle, 163*, 35–75.

Rikap, C. (2021). *Capitalism, power and innovation: Intellectual monopoly capitalism uncovered*. Routledge.

Sampat, B. N. (2018). *A survey of empirical evidence on patents and innovation*. National Bureau of Economic Research.

Schumpeter, J. A. (1942). *Capitalism, socialism and democracy*. Routledge.

Schwartz, H. M. (2016). Wealth and secular stagnation: The role of industrial organization and intellectual property rights. *The Russell Sage Foundation Journal of the Social Sciences, 2*(6), 226–249.

Schwartz, H. M. (2020). Intellectual property, technorents and the labour share of production. *Competition & Change*, 1024529420968221.

Shapiro, C. (2018). Antitrust in a time of populism. *International Journal of Industrial Organization, 61*, 714–748.

Srnicek, N. (2017). *Platform capitalism*. Wiley.

Summers, L. H. (2014). Reflections on the 'new secular stagnation hypothesis'. *Secular Stagnation: Facts, Causes and Cures*, 27–38.

Summers, L. H. (2016). The age of secular stagnation: What it is and what to do about it. *Foreign Affairs, 95*(2), 2–9.

Teixeira, R. A., & Rotta, T. N. (2012). Valueless knowledge-commodities and financialization: Productive and financial dimensions of capital autonomization. *Review of Radical Political Economics, 44*(4), 448–467.

UNCTAD. (2019). *Digital economy report 2019: Value creation and capture—Implications for developing countries*. United Nations.

US Congress. Subcommittee on antitrust, commercial and administrative law. (2020). *INVESTIGATIONOF COMPETITION INDIGITALMARKETS*. US Congress.

Veblen, T. (1908). On the nature of capital: Investment, intangible assets, and the pecuniary magnate. *The Quarterly Journal of Economics, 23*(1), 104–136.

WIPO. (2017). *World intellectual property report 2017: Intangible capital in global value chains*. World Intellectual Property Organization.

World Intellectual Property Organization. (2019). *WIPO Technology trends 2019: Artificial intelligence*. WIPO.

CHAPTER 3

Tech Giants' Corporate Innovation Systems

Abstract In this chapter, we show that tech giants organize what we define as *corporate innovation systems* from which they appropriate knowledge that they transform into intangible assets garnering intellectual rents. We define corporate innovation systems as systems that are organized and controlled by an intellectual monopoly and include a multitude of more or less subordinate organizations participating in production and innovation networks. The chapter studies selected US and Chinese tech giants' corporate innovation systems through an analysis of their scientific publications' co-authoring organizations, their participation in the open-source software environment and start-up acquisitions.

Keywords Tech giants · Corporate innovation systems ·
Open-source software · Technology acquisitions · Knowledge predation

3.1 Introduction

In Chapter 2, we analysed the formation of intellectual monopolies and the mechanisms they use to privatize knowledge and to transform knowledge into intangible assets. In this chapter, the focus is on how they harvest knowledge from outside their organizations. Tech giants combine internal knowledge creation with accessing knowledge and data from other organizations worldwide. To capture these new forms of organization where continuous innovation is organized and controlled by the intellectual monopoly, we propose a revised version of the concept *corporate innovation system* (CIS) (Granstrand, 2000). By establishing a CIS constituted by a multitude of more or less subordinate firms and knowledge institutions, including linkages to open-source software communities, these companies reproduce and reinforce their positions as intellectual monopolies.

In Sects. 3.2 and 3.3 we introduce the CIS concept and explain why there is a need for such a new concept. Sects. 3.4 and 3.5 show how intellectual monopolies from the US and China organize their CIS by outsourcing knowledge production and hereby harvest knowledge from universities and open-source software (OSS) communities. Section 3.6 elaborates on acquisitions as a means to constantly reconfigure tech giants' CIS and Sect. 3.7 offers preliminary conclusions on the consequences of knowledge predation by intellectual monopolies.

3.2   The Corporate Innovation System

To rework on Granstrand's (2000) almost forgotten concept "Corporate Innovation Systems", we combine insights from the literature on innovation systems and the literature on global value chains. The former contributes with its focus on innovation as an interactive process, while the latter attributes a crucial role to the (rent-seeking) lead firm in the organization of networks of actors.

There are less than a handful of academic contributions that make use of the CIS concept and most of those define it as referring to the organization of in-house activities. An interesting example, close to our interest, is the analysis of Google as a CIS in Steiber and Alänge (2013). The authors do take into account that the system is open, but they do not include external organizations in it. This stands in contrast to Granstrand's (2000, p. 13) original definition:

> A '**corporate innovation system**' is the set of actors, activities, resources and institutions and the causal interrelations that are in some sense important for the innovative performance of a corporation or groups of collaborating companies and other actors (e.g. universities, institutes, agencies).

Here, we take this definition as starting point, suggesting three important revisions. First, we focus on corporate systems organized around a single firm—disregarding those organized around a group of firms. Second, we add a focus on vertical dominance by the firm around which the system is established. As we will show next, intellectual monopolies simultaneously control and orient their CIS in order to preserve their dominance.[1] Third, we focus on CIS operating on a global scale. The reason for these revisions is that we are primarily interested in data-driven intellectual monopolies. These corporations establish their dominance worldwide through data centralization—in part thanks to their control of platforms—and by constant innovations on how to process and analyse big data (see Chapter 2 on degrees of intellectual monopoly).

Regarding our second revision, we emphasize that by organizing CIS, intellectual monopolies provide the general orientations and desired results to other participants, of course without being able to anticipate every step to be followed. With some relative autonomy for the other participants, the final outcomes remain uncertain, but leaders divert some of the associated economic risk to other participants (Rikap, 2018). Serendipity or haphazard innovations in newcomer firms are still feasible, but in most cases, the aspiration is not to challenge the dominance of the tech giant but rather to sell new ideas or even selling the new enterprise to the intellectual monopoly controlling the CIS.

3.3   Why Do We Need a New Concept? Revisiting the Literature

Certainly, the acknowledgement of a central planning and organizing capacity of leader corporations is not new.

In the 1990s, the flagship firm theory focused on how multinational corporations organized and set the strategy of multiple value chains (Rugman & D'Cruz, 1997). However, as Strange and Humphrey (2018) note, this strategic position of the flagship firm does not lead Rugman and D'Cruz (1997) to account for the asymmetric power and inequal value appropriation resulting from the capacity (that they do recognize) of the flagship firm to set the strategy of the network.

---

[1] Of course, the concept CIS could as well be applied to a much wider set of corporations (even to small local firms engaged in innovation and integrated in local or national networks).

The recognition of unequal inter-firm relations organized by lead corporations has been a major contribution of the Global Value Chain (GVC) literature. Global value chains are presented as a production process sliced in pieces that take place in different firms and geographies but that is organized and governed by a lead corporation (Gereffi, 2014; Gereffi et al., 1994, 2005; Ponte & Sturgeon, 2014; Sturgeon, 2009). Hierarchical relations between leader firms and partners and contestation over value creation and distribution are considered as explicit possibilities, and leader corporations can exercise behaviour control due to power asymmetries reflecting uneven access to strategic markets and/or technologies (Gereffi et al., 2005; Strange & Humphrey, 2018). Power is exercised directly in a one-to-one relationship and also in more diffuse ways, such as by setting industry standards (Dallas et al., 2017).

Most of the GVC analysis refers to transactions in supply chains in relation to tangible products and innovation processes are not in focus (Chaminade et al., 2016; Jurowetzki et al., 2018). This literature generally assumes that knowledge-intensive activities remain internalized by the leading firm, while manufacturing and other non-core activities are generally outsourced (or offshored) (Serfati, 2008). Likewise, within the flagship firm approach, Rugman and D'Cruz (1997) only state that the direction of the overall innovation process will be set by the flagship firm. There is no analysis of the actual innovation process and how it is organized.

To some extent, this gap has been filled by suggestions for alternative structures, such as Global Innovation Networks (GIN) (Chaminade et al., 2016; Ernst, 2008, 2009; Liu et al., 2013; Parrilli et al., 2013) and innovation and platform ecosystems (Adner & Kapoor, 2016; Cennamo & Santalo, 2013; Gawer & Cusumano, 2014; Jacobides et al., 2018; Stallkamp & Schotter, 2019). Unlike GVC, GIN literature is ambivalent when it comes to the role of leading corporations. Ernst (2008, 2009), exceptionally, takes a clear stance in this respect: GIN are hierarchical structures organized by leader corporations. He thinks of power relations in the GIN as equivalent to those in global production networks and GVC literature. Other authors consider GIN as structures not necessarily subjected to hierarchical relations (Chaminade et al., 2016; Liu et al., 2013).

In the case of platforms, and more broadly the ecosystem literature, most contributions have overlooked power asymmetries (Chaminade et al., 2018; Jurowetzki et al., 2018; Lundvall, 2002) and knowledge predation. Ecosystems are defined as structures integrated by firms with unique or supermodular complementarities (Cennamo & Santalo, 2013; Gawer & Cusumano, 2014; Jacobides et al., 2018; Stallkamp & Schotter, 2019). They are seen a constituted by interacting organizations *not hierarchically managed* (Jacobides et al., 2018). Albeit useful for analysing those complementarities, authors overlook the distribution of value and knowledge among participating actors. Furthermore, tech giants organize multiple interrelated platforms going beyond organizing a single platform ecosystem. They also establish traditional collaborations with universities and other research institutions not based on such multilateral and non-generic complementarities.

Overall, existing theories (such as flagship, GVC, innovation network and ecosystems) grasp different parts, but not all relevant aspects of corporate leaders organizing innovation beyond their legally bound capital. As we argued in Chapter 2, intellectual monopolies appropriate value not only from their production networks, platforms and GVC, but also by garnering intellectual rents from innovative activities taking place in other organizations.

Other organizations accept subordination because this is their best survival strategy, but this does not mean that the relationship is equally beneficial. In fact, this unequal relation can be considered as a form of predation. Veblen (2017) conceptualized predation as a direct manifestation of superior force. Mobilizing this concept in the intellectual monopoly context, Rikap (2020) reconceived predation as a direct production relation of spoliation where a corporation exercises its superiority by planning other organizations' activities. Different authors have provided evidence of leaders' capacity to appropriate value within production networks (Bergvall-Kareborn & Howcroft, 2013; Kraemer et al., 2011; Selwyn, 2019; Smith, 2016). What has been overlooked, is that leaders appropriate knowledge also from outsourced innovation activities that they transform into intangible assets. All in all, we need a concept that simultaneously refers to how these corporations organize and plan innovation and to how they use their leading position to extract intellectual rents not only from their in-house R&D but also from the successful results achieved in their outsourced innovation activities.

3.4   OUTSOURCING R&D

The evolution and performance of the CIS reflect internal efforts as well as access to external knowledge. Internal innovation efforts include investments in R&D, training of employees and building learning organizations. They also encompass the formation of new capabilities attained through hiring. Innovation activities reflect the combination of science-based and experience-based knowledge. In this section, the focus is on science-based knowledge and on the formation of external linkages related to research.

Figure 3.1 shows the centrality given by selected tech giants to R&D investments. They are among the top corporations in business expenditure on R&D (BERD) and sustain high R&D investments as their revenues keep rising. The cases of Amazon—the world's first company in BERD—and Alibaba (29th globally) are particularly impressive in this respect. Amazon also stated in its 2019 annual report that it would spend USD 700 million (which represents 6% of its 2019 net income) to provide access to training programs in high-demand fields (including healthcare, cloud computing and machine learning) to more than 100,000 employees (Amazon, 2020a).

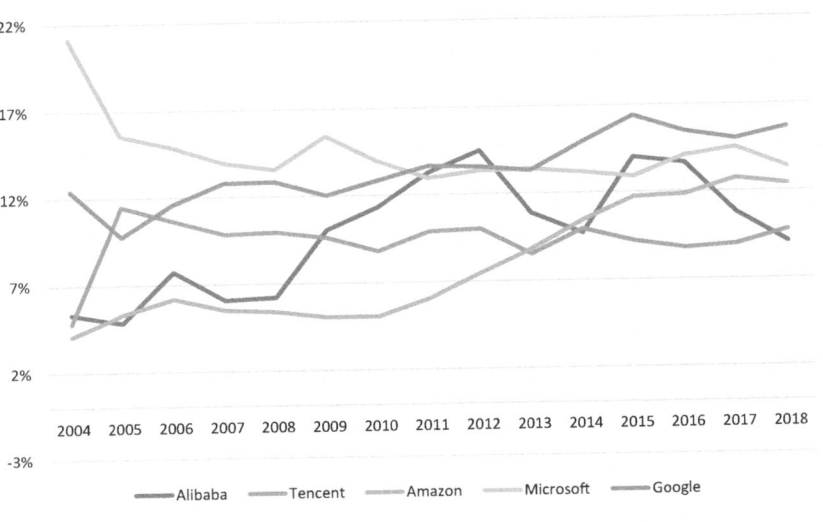

Fig. 3.1   R&D over revenues. Selected tech giants (*Source* Compustat)

Externally, the institutional context in terms of regional and national innovation systems conditions the evolution and performance of corporate innovation systems, and innovation hubs offer intellectual monopolies a chance to profit from concentrated resources and capabilities (Rikap & Flacher, 2020). According to WIPO, big shares of innovation inputs and outputs are concentrated in the US and China (Dutta et al., 2019) and within these countries digital innovation takes place in a handful of innovation hubs.

Balland et al. (2020) have found that innovation remains increasingly concentrated in a few cities in the US. Among them, the Silicon Valley stands out. In the case of the digital economy, another major innovation hub is Seattle (home of Amazon and Microsoft's headquarters). There are only a few other innovation hubs in the US located on the East Coast (primarily New York and Boston). Likewise, in China innovation is geographically concentrated and limited to a few cities. Shenzhen stands out and has been compared to the Silicon Valley (Chen & Ogan, 2017). Beijing and Shanghai are the other two main innovation hubs for the digital economy.

It is important to note that Google has R&D locations in these three Chinese cities, while Microsoft is based in the latter two. Microsoft founded its Microsoft China Research Institute in Beijing in 1998, and in 2010 it opened the Shanghai High-tech phase one campus, the first Microsoft R&D and service campus outside the US.

One way to illustrate the predatory character of intellectual monopolies is to contrast the high frequency of co-publication with their low frequency of co-patenting. Table 3.1 presents this comparison for selected tech giants from the US and China.

Altogether, publications' co-authorship and patents' co-ownership indicate that these corporations garner intellectual rents from their CIS not shared with their co-authors, thus monetizing knowledge at the expense of those organizations' R&D (Table 3.1) (see Chapter 5 for evidence on the alignment of publications' and patents' contents). Building on Durand and Milberg's (2020) typology, one can argue that tech giants combine data-driven innovation rents with legal monopoly rents. This is an extended practice in AI research. None of the top 20 patent applicants co-owns more than 1% of its AI portfolio (World Intellectual Property Organization, 2019).

Table 3.1 Selected tech giants' publications and patents, co-authorships and co-ownerships

Company	Publications (until 2019 included)	Co-authoring organizations	Applied and granted patents (until 2017 included)	Co-owned patents with other organizations
Amazon	824	766	10,063	13 (0.1%)
Microsoft	17,405	4025	76,109	160 (0.2%)
Google	6447	3397	25,538	65 (0.3%)
Tencent	643	366	5462	13 (0.2%)
Alibaba	685	427	3532	0 (0%)

Source Web of Science and Derwent innovation (this database includes all the information from the following patent offices: USPTO, WIPO, European, Japan, Australian, British, Canadian, French, German, Russian and Korean patent offices)

These results are in line with the examples presented by Popkin (2019). The author interviewed academic researchers receiving funding from tech giants. He observed that these companies generally launch an open call for projects once a year, usually providing grants that are small but still meaningful for academic researchers. Amazon's last call received over 800 applications but funded only 49. Contacts with employees, such as previous professional connections or networking at conferences, are important for accessing funds from tech giants. They are even a precondition for receiving a grant from Microsoft. In his interviews, funded researchers pointed out that companies avoid discussing new products with them. They are not usually included when it comes to sharing profits of a product based on their research. Furthermore, tech giants do not "provide access to any internal or proprietary data" (Popkin, 2019, p. 666).

These types of calls for projects save time for tech giants, because researchers apply themselves, thus there is no need to seek out who is working on what. They also save money because providing small grants to researchers is cheaper than expanding in-house R&D teams, which not only implies paying for the research project itself, but also for the people employed. Additionally, this mechanism assures that the tech giant has access to research covering a wide diversity of fields. Sustaining themselves as multi-technology and multi-product companies would (as we show in Chapter 4) be way more expensive if done in-house.

Beyond funding academic researchers, another mechanism that enables collaborations is hiring academic researchers that keep ties with former colleagues. For instance, in 2015, the US top mental health researcher left the National Institute of Mental Health to work for Alphabet (Google).[2] Tech companies can offer scholars access to unique data sets as well as attractive salaries, which are major reasons why a number of scholars in the US have changed their path from public life sciences departments to work for tech giants.[3]

Table 3.2 presents each company's top 10 co-authors between 2014 and 2019. It provides evidence of technological cooperation among tech giants such as between Google, Amazon and Microsoft, as well as with IBM for the cases of Amazon and Google, and between Alibaba and Microsoft. Google, Amazon and Microsoft share some of the same US universities as most frequent co-authors, with the University of California always ranked first. Likewise, Tencent and Alibaba rely on the same Chinese universities, the Chinese Academy of Science and a Singaporean university.

Microsoft has the most diverse network of privileged partners. It includes two Chinese and two European universities. In fact, 15 of its top 100 co-authors in this period were Chinese organizations and three universities from Hong Kong. Among these collaborations, one that particularly raised the US government concern was Microsoft's joint AI research with China's National University of Defense Technology, which operates under the auspice of China's Central Military Commission.[4]

As can be seen from Table 3.2, the most frequent co-authors come from the corresponding corporation's home country. At first glance, it indicates that the tech giants primarily draw upon the science base of their own national innovation system. They benefit from privileged access to knowledge resources financed by taxpayers in their own country—but without sharing the resulting intellectual rents.

But the tech giants' scientific collaborations combine domestic with foreign partners. We have already referred to the importance of Chinese organizations for Microsoft's CIS. Moreover, despite the centrality of US

---

[2] Retrieved from https://www.technologyreview.com/2015/09/21/10487/why-americas-top-mental-health-researcher-joined-alphabet/ last access October 5, 2020.

[3] Retrieved from https://www.nature.com/news/why-biomedical-superstars-are-signing-on-with-google-1.18600 last access October 3, 2020.

[4] https://www.ft.com/content/9378e7ee-5ae6-11e9-9dde-7aedca0a081a.

Table 3.2 Selected tech giants' top co-authors (2014–2019)

Microsoft	Google	Amazon	Tencent	Alibaba
University of California	University of California	University of California	Chinese acad of science	University of Science and Technology of China
University of Washington	Stanford University	Microsoft	Chinese University of Hong Kong	Chinese acad of science
University of Sci & Tech of China	Microsoft	University of Washington	Tsinghua University	Zhejiang University
MIT	MIT	Google	Peking University	Tsinghua University
Tsinghua University	Harvard	IBM	Shanghai Jiao Tong University	Peking University
University of London	Carnegie Mellon University	Georgia Inst of Technology	Sun Yat-sen University	Shanghai Jiao Tong University
Carnegie Mellon University	University of Illinois	Carnegie Mellon University	Zhejiang University	Microsoft
Google	University of Washington	University of Texas	Harbin Institute of Technology	Fudan University
Stanford University	IBM	MIT	Beihang University	Nanyang Technological University
ETH Zurich	New York University	Indian Inst of Technology	Nanyang Technological University	Nanjing University

Source Authors' analysis based on Web of Science

collaborators, Amazon co-authored 47% (8.7%) of its papers with at least one non-US (Chinese) based institution. The two Chinese corporations have even higher shares of transnational collaborations. 54% (37%) of Alibaba's and 47% (30%) of Tencent's publications were co-authored with at least one non-Chinese (a US) based organization. After Microsoft, IBM and Tencent follow as Alibaba's most frequent corporate co-authors (positions 12 and 26, respectively). Baidu is Tencent's most frequent corporate co-author and Microsoft follows (22nd and 26th position, respectively).[5]

[5] Moreover, Alibaba's and Tencent's executives have more transnational than national corporate ties. These transnational ties vis-à-vis total corporate ties are—in percentage—greater than those of US big corporations' board members (de Graaff, 2020).

Additionally, only two of Alibaba's new AI centres (called DAMO Academy) are in China. There are three in the US (Seattle, Sunnyvale and New York), one in Israel and one in Singapore.[6] Alibaba claims that the program "aims to integrate science with industry and speed up information exchange" (Alibaba, 2019). These centres are near leading research universities, looking both to profit from their research capabilities and attract talent.[7] Tencent has also opened an artificial intelligence research centre in Seattle.[8]

3.5 Profiting from Open-Source Software and App Developers

Within their respective CIS, Google, Amazon, Microsoft, Alibaba and Tencent sometimes choose to organize open-source innovation networks and outsource applications' development.

When opening their platforms to developers, tech giants define the conditions (fees, which apps end up being offered, the software to program with, etc.) without paying direct remuneration to developers; and the developers take the risk and cover the costs (Bergvall-Kareborn & Howcroft, 2011, 2013; Rikap, 2018). Google's Android is an open-source platform for mobile phone development. Google offers it for free to handset makers but requires "Google search" installed as a default setting, which the European Commission found was an illegal practice strengthening Google's search engine dominance.[9] Android is used by 74.6% of the world's smartphones (24.8% runs on iOS, Apple's mobile operating system) and, by August 2020, 3.3 million applications were available on GooglePlay.[10]

Besides Google (and Apple), Amazon also outsources app development for its virtual assistant, with more than 700,000 developers building Alexa skills by August 2020 (Amazon, 2020b). In the case of Chinese giants,

[6] https://damo.alibaba.com/about/.

[7] https://techcrunch.com/2017/10/10/alibaba-group-will-invest-15b-into-a-new-global-research-and-development-program/.

[8] https://techcrunch.com/2017/04/28/tencent-to-open-ai-research-center-in-seattle/?_ga=2.127708877.1164454910.1592757734-700995395.1592226319.

[9] https://ec.europa.eu/competition/elojade/isef/case_details.cfm?proc_code=1_40099.

[10] https://www.businessofapps.com/guide/app-stores-list/.

to promote digital content and entertainment creators, Tencent created a specialized incubator called Tencent WeStart. By training start-ups, Tencent expands its CIS, integrating an increasing number of subordinate innovating firms.[11]

Overall, developers do not contribute to the core infrastructure of tech giants' platforms. They create complementary products but partially profit from them (there is always a fee charged by the platform owners for using the platform).

The inequal distribution of the value extracted by engaging in open-source projects is not as easy to evidence as outsourced applications. The result of an OSS project is available to everyone. However, due to the high degree of interconnectivity of the digital industry, those enterprises which can integrate it into more complex systems and those with access to big datasets have much better chances to monetize results. This feature of digital (shared to some extent with other ICT) technologies favours big tech companies who keep their most sensitive technologies and datasets secret.

Microsoft's acquisition of GitHub for USD 7.5 billion[12] in 2018, a leading software development platform for developers, evidenced a turning point in its position regarding open source. Hereby it accessed professional data of millions of developers. By October 2018, Github's top three open-source projects, ranked by number of participating developers, belonged to Microsoft, Facebook and Google: Microsoft/vscode (19,000 developers), facebook/react-native (10,000 developers), tensorflow/tensorflow (9300 developers). These examples evidence that tech giants are profiting from developers far beyond their own workforce because Microsoft had, by the same date, 7700 registered developers, Google 5500 and Facebook 1700 on Github (https://octoverse.github.com/2018/).

In turn, Alibaba states that it is active in the OSS community, with contributions to more than 100 projects by March 2017 (Alibaba, 2018). Its open-source projects in GitHub were, by 2020, the most active among those put in open source by Chinese enterprises (X-lab, 2020). A similar number of projects had been open sourced by Tencent by early 2020

---

[11] https://www.westarthk.com/.

[12] https://news.microsoft.com/2018/06/04/microsoft-to-acquire-github-for-7-5-billion/.

on topics like cloud computing, big data, AI healthcare and network security.[13]

Overall, the US and Chinese tech giants profit from open-source innovation in multiple and complementary ways:

1. Accessing reusable code for their private innovations. For instance, Microsoft included a Linux kernel in Windows 10 so that it can now profit from Linux developments, in part withdrawn from GitHub.[14]
2. They profit from developers' free work (Schrape, 2018).
3. Open-source activities contribute with valuable information on the type of recurrent problems faced by smaller companies, and offer potential users' feedback in advance, thereby providing valuable inputs for cloud services.
4. By releasing libraries to open source (such as Tensor Flow and PyTorch), standards are set, and thus small companies can be easily integrated into their business in case they are acquired.

All in all, while open CIS favour collaborative knowledge production, most of the resulting gains are collected by corporate leaders, reinforcing their intellectual monopolies. This and the previous section indicate that the openness of tech giants' CIS is uni-directional. While the intellectual monopoly is open for incoming knowledge, it is quite closed when it comes to sharing protected and secret knowledge and especially knowledge related to core competencies.

3.6   ACQUIRING TECHNOLOGY AND ENTERING MARKETS THROUGH ACQUISITIONS OR INVESTING IN SMALLER COMPANIES

Tech giants' CIS are in constant flux, reflecting high rates of capital accumulation and innovation. A central mechanism to speed up the building of technological capabilities in emerging new technologies without engaging in the associated risks is the acquisition of promising

---

[13] https://www.tencent.com/en-us/articles/2201019.html.

[14] Retrieved from https://www.theverge.com/2019/5/6/18534687/microsoft-windows-10-linux-kernel-feature last access December 21, 2019.

start-ups. Google, Apple and Microsoft as well as Alibaba have used this strategy when it comes to AI.

These acquisitions create a "kill zone", as Kamepalli et al. (2020) showed for the cases of companies acquired by Google and Facebook. When one of these giants acquires a start-up, venture capitals reduce their investments (in number of deals and invested amounts) in competing companies or companies in close markets. This works as a form of self-fulfilling prophecy: venture capitals limit investments anticipating that the acquisition would lead to a winner take all market. This contributes to explaining the slowdown in the US entrepreneurship rate since the 2000s.[15]

According to WIPO (2019), Alphabet (Google) ranks first in AI-related acquisitions. Since 2009 it has acquired 18 AI companies, including the AI forerunner UK start-up DeepMind in 2014. Apple and Microsoft followed (11 and 9 AI acquisitions until May 2018).

Since 2010, when Microsoft launched Azure, it has been actively purchasing both cloud computing and mobile platform firms (15 and 8 until 2016 included) (Lopez Giron & Vialle, 2017). Cloud-related acquisitions dealt first with internalizing R&D capabilities and, since 2014, they are focused on AI. Cloudyn (for managing cloud costs) and Bonsai (an AI training platform) are illustrative examples. Acquiring large patent portfolios through acquisitions in AI was observed by WIPO (2019) as a particular characteristic of Microsoft.

The public cloud is a permanent source of data providing insights into which companies to acquire. Amazon, Microsoft, Google and Alibaba are the four leading companies in market share of the cloud computing business. They can easily detect business areas where consumption of web services suddenly bumped, thus indicating potentially promising businesses and technologies.

For instance, through Amazon Web Services (AWS), Amazon spotted and gave seed money to Yieldex, Sonian, Engine Yard and Animoto.[16] Examples of Amazon digital technology acquisitions include

---

[15] Haltiwanger et al. (2014) showed that the number of young firms (aged up to five years) in the US in the high-tech sector is declining since 2000, while for the overall private sector this figure remains relatively flat until 2008. Since then, the fall is steeper for high-tech than for the overall private sector.

[16] Retrieved from https://www.reuters.com/article/amazon-cloud-idUSN1E7A727 Q20111109 last access February 27, 2019.

the pioneering cloud start-up Amiato inc., Audible inc. and the robotic start-up Kiva Systems inc. At least until 2017 included, Audible's was the biggest patent portfolio acquired by Amazon. Amazon paid USD 1.075 billion for acquiring Audible and Kiva. Amazon's acquisitions also include six AI companies (World Intellectual Property Organization, 2019). Among them, TSO Logic was a Canadian start-up that retrieves and analyses company data to build a granular statistical model to calculate the cost savings in connection with company transitions to the cloud. In 2019, Amazon also acquired Canvas Technology, specialized in robots that autonomously navigate warehouses. These acquisitions signal Amazon's innovation priorities also aligned with its recent publishing trends, as we show in Chapter 5.

Chinese data-driven intellectual monopolies have also acquired companies, including firms operating in big potential markets like East Asia and Brazil.[17] Jia et al. (2018) argue that purchasing firms and equity stakes have been Alibaba and Tencent's main globalization strategy. Moreover, Chinese giants also invest in companies for their technology. Between 2014 and 2018, Baidu, Alibaba and Tencent took equity stakes in 39 AI start-ups, 17 coming from the US (CBInsights, 2018). Overall, by mid-2021, Alibaba had acquired 34 companies and invested in 227 and Tencent had acquired 18 companies, but invested in 582.[18]

A prominent example was Tencent's equity stake in Suiyuan Technology, an AI chip maker (Ernst, 2020). Significant investments include Tencent purchased equity in Snapchat and, in 2018, an equity-swap with Spotify.[19] While these investments seem strategic, Jia et al. (2018) consider that others, like investments in ride-hailing companies, were financially driven. Since the Covid-19 pandemic, Tencent has taken advantage of some foreign companies' falling valuation and acquired Funcom, a Norwegian game developer. It also took a stake in Yager, a German developer, and poured capital into multiple fintech start-ups.[20]

In the same vein, Alibaba has pursued an equity investment strategy, sometimes motivated by getting access to advanced technologies, such as

[17] Retrieved from https://www.economist.com/business/2018/07/07/chinese-and-us-tech-giants-go-at-it-in-emerging-markets last access April 19, 2020.

[18] https://www.crunchbase.com/organization/tencent and https://www.crunchbase.com/organization/alibaba.

[19] https://techcrunch.com/2018/02/28/tencent-music-spotify/.

[20] https://www.ft.com/content/844ed28c-8074-4856-bde0-20f3bf4cd8f0.

the encryption technology firm V-Key and the Israeli company ThetaRay, specialized in financial network security. This strategy also included firms from the US, Canada and India (Jia et al., 2018). Among Alibaba's recent purchases, Tango had developed technologies to adapt instant messaging for e-commerce, and MagicLeap was specialized in AR/VR technologies. In 2019, Alibaba also acquired a Chinese AI chip designer, C-Sky Microsystems (Ernst, 2020).

Unlike Alibaba and Tencent, GAFAM have generally preferred acquisitions over equity investments. Nonetheless, Facebook recently invested $5.7bn (almost 10% stake) in Reliance Jio, an Indian telecoms company. A Financial Times article claimed that Facebook aimed to create *a super app* for India, like Tencent's "WeChat".[21]

Investing in equity stakes may be preferred to acquisitions. Alibaba, Tencent and now Facebook can still plan and organize smaller companies and get access to their data and technology when they invest in such equity stakes. Hence, they get similar benefits as from acquisitions but with lower associated risks. Moreover, considering economic risks and the threat of antitrust regulations, when companies have established technological or market leadership, tech giants may prefer to keep them formally independent but still under their influence.

Through acquisitions the intellectual monopoly reconfigures its CIS adapting it to its strategic objectives. It does so on the basis of a combination of technological capabilities and financial strength (particularly through using idle liquid assets in tax havens) (Bryan et al., 2017; Pozsar, 2018). In some cases, the acquisition aims at getting rid of a potential competitor, while in other cases the aim is to speed up technological advances in what the leading corporation has defined as strategic technologies, such as AI. As a result, the acquisitions reinforce the scale and scope advantage of the intellectual monopoly. The CIS expands both in geographical space and in its mastering of diverse technologies. An acquisition means not only the purchase of a company but also access to the acquired firm's innovation network.

---

21 https://www.ft.com/content/6513ca1a-ee3b-4bd1-a975-a154043de557.

3.7 Conclusions

This chapter has introduced the concept Corporate Innovation System and applied it to selected tech giants. Their CIS combine the exploitation of local and national innovation systems with harvesting knowledge globally. While led by a private for-profit corporation, global CIS encompass not only private knowledge production, but integrate also public sector organizations and open science initiatives. Intellectual monopolies have the capacity to transform knowledge produced in these contexts into intangible assets, thus ultimately extracting rents from public knowledge. This emphasis on asymmetric, predatory relations in the innovation process—within and beyond platforms—distinguishes the CIS from other concepts dealing with open innovation relations.

Intellectual monopolies act like venture capitals for multiple start-ups. They follow them closely and keep them within their CIS for as long as they promise profitable businesses. Acquisitions aiming to get access to the acquired company's technology differ from acquisitions aiming to increase the market share of the leader firm. The former expands market power while the latter extends intellectual monopoly to wider fields of technology. In both cases, other venture capital investments are discouraged, leading to an overall lower entrepreneurship rate in high-tech.

For certain steps of different innovation processes—typically general-purpose technology innovations—this chapter also provided evidence of technological cooperation among the tech giants (see Table 3.2). In these cases, global CIS belonging to different intellectual monopolies intersect. Technological cooperation between intellectual monopolies makes it possible to undertake big and costly R&D projects (Rikap, 2019).

The formation and reconfiguration of global CIS transform the geography of global knowledge production. Our analysis points to how tech giants' global outreach to customers, suppliers and users is combined with highly localized interaction within and between their distinct CIS. In both the US and China there is a significant geographical concentration of tech giants' R&D facilities in a handful of locations. This high degree of concentration is reflected in the geographical patterns of networking within CIS. Not only do we see a pattern where the leading company has the most intense collaboration with nearby knowledge institutions, we also see how Chinese companies locate R&D activities in the two main

US high-tech hotspots and how the US companies locate such activities in the corresponding Chinese hotspots.

Tech giants dominate their CIS through their command of digital technologies and for the five selected enterprises through their strength in AI (see Chapter 4). They are constantly thriving to monopolize the control of AI as a general-purpose technology with the potential to become a general-purpose method of invention, aiming at becoming intellectual monopolies of fourth degree (see Chapter 2). In the next chapter, we explore further the key role AI plays for these selected tech giants as well as the latter's crucial positioning in defining the AI technology innovation system.

References

Adner, R., & Kapoor, R. (2016). Innovation ecosystems and the pace of substitution: Re-examining technology S-curves. *Strategic Management Journal, 37*(4), 625–648.

Alibaba. (2018). *Annual report 2017*. Alibaba.

Alibaba. (2019). *Annual report 2018*. Alibaba.

Amazon. (2020a). *Amazon annual report 2019*. Amazon.

Amazon. (2020b). *Small business success in challenging times: 2020 AMAZON SMB IMPACT REPORT*. Amazon.

Balland, P.-A., Jara-Figueroa, C., Petralia, S. G., Steijn, M. P., Rigby, D. L., & Hidalgo, C. A. (2020). Complex economic activities concentrate in large cities. *Nature Human Behaviour, 4*(3), 248–254.

Bergvall-Kareborn, B., & Howcroft, D. (2011). Mobile applications development on Apple and Google platforms. *Communications of the Association for Information Systems, 29*(1), 565–580.

Bergvall-Kareborn, B., & Howcroft, D. (2013). The Apple business model: Crowdsourcing mobile applications. *Accounting Forum, 37*, 280–289.

Bryan, D., Rafferty, M., & Wigan, D. (2017). Capital unchained: Finance, intangible assets and the double life of capital in the offshore world. *Review of International Political Economy, 24*(1), 56–86.

CBInsights. (2018, April). *Rise of China's big tech in AI: What Baidu, Alibaba, and Tencent are working on*.

Cennamo, C., & Santalo, J. (2013). Platform competition: Strategic trade-offs in platform markets. *Strategic Management Journal, 34*(11), 1331–1350.

Chaminade, C., De Fuentes, C., Harirchi, G., & Plechero, M. (2016). The geography and structure of global innovation networks: Global scope and regional embeddedness. In *Handbook on the geographies of innovation*. Edward Elgar Publishing.

Chaminade, C., Lundvall, B. A., & Haneef, S. (2018). *Advanced introduction to national innovation systems*. Edward Elgar Publishing.

Chen, X., & Ogan, T. L. (2017). China's emerging Silicon Valley: How and why has Shenzhen become a global innovation centre. *European Financial Review*, 55–62.

Dallas, M., Ponte, S., & Sturgeon, T. (2017). *A typology of power in global value chains*. Copenhagen Business School.

de Graaff, N. (2020). China Inc. Goes global. Transnational and national networks of China's globalizing business elite. *Review of International Political Economy, 27*(2), 208–233.

Durand, C., & Milberg, W. (2020). Intellectual monopoly in global value chains. *Review of International Political Economy, 27*(2), 404–429. https://doi.org/10.1080/09692290.2019.1660703

Dutta, S., Lanvin, B., & Wunsch-Vincent, S. (2019). *The global innovation index 2019: Creating healthy lives—The future of medical innovation*. World Intellectual Property Organization (WIPO), Geneva, Switzerland.

Ernst, D. (2008). *Can Chinese IT firms develop innovative capabilities within global knowledge networks?* https://papers.ssrn.com/sol3/papers.cfm?abstract_id=2742950

Ernst, D. (2009). *A new geography of knowledge in the electronics industry? Asia's role in global innovation networks*. https://papers.ssrn.com/sol3/papers.cfm?abstract_id=2742923

Ernst, D. (2020). *Competing in artificial intelligence chips: China's challenge amid technology war*. Centre for International Governance Innovation.

Gawer, A., & Cusumano, M. A. (2014). Industry platforms and ecosystem innovation. *Journal of Product Innovation Management, 31*(3), 417–433.

Gereffi, G. (2014). Global value chains in a post-Washington consensus world. *Review of International Political Economy, 21*(1), 9–37.

Gereffi, G., Humphrey, J., & Sturgeon, T. (2005). The governance of global value chains. *Review of International Political Economy, 12*(1), 78–104.

Gereffi, G., Korzeniewicz, M., & Korzeniewicz, R. P. (1994). Global commodity chains. In *Commodity chains and global capitalism* (pp. 1–14).

Granstrand, O. (2000). *Corporate innovation systems: A comparative study of multi-technology corporations in Japan*. Chalmers University, Gothenburg.

Haltiwanger, J., Hathaway, I., & Miranda, J. (2014). *Declining business dynamism in the US high-technology sector*. Available at SSRN 2397310.

Jacobides, M. G., Cennamo, C., & Gawer, A. (2018). Towards a theory of ecosystems. *Strategic Management Journal, 39*(8), 2255–2276.

Jia, K., Kenney, M., & Zysman, J. (2018). Global competitors? Mapping the internationalization strategies of Chinese digital platform firms. In R. Tulder, A. Verbeke, & L. Piscitello (Eds.), *International business in the information and digital age* (pp. 187–216). Emerald Publishing Limited.

Jurowetzki, R., Lema, R., & Lundvall, B.Å. (2018). Combining innovation systems and global value chains for development: Towards a research agenda. *The European Journal of Development Research, 30*(3), 364–388.

Kamepalli, S. K., Rajan, R., & Zingales, L. (2020). *Kill zone*. National Bureau of Economic Research.

Kraemer, K., Linden, G., & Dedrick, J. (2011). *Capturing value in Global Networks: Apple's iPad and iPhone*. University of California, Irvine, University of California, Berkeley, y Syracuse University, NY. Http://Pcic.Merage.Uci.Edu/Papers/2011/Value_iPad_iPhone.Pdf. *Consultado El, 15.* https://pdfs.semanticscholar.org/9cb5/262a46e7c9131de43433b7c5f9b65386f8e2.pdf

Liu, J., Chaminade, C., & Asheim, B. (2013). The geography and structure of global innovation networks: A knowledge base perspective. *European Planning Studies, 21*(9), 1456–1473.

Lopez Giron, A. J., & Vialle, P. (2017). A preliminary analysis of mergers and acquisitions by Microsoft from 1992 to 2016: A resource and competence perspective. In *28th European Regional Conference of the International Telecommunications Society (ITS): 'Competition and Regulation in the Information Age'*, Passau, Germany.

Lundvall, B. A. (2002). *The university in the learning economy*. Presentation on the Future Role of Universities, Strasbourg, 26.

Parrilli, M. D., Nadvi, K., & Yeung, H. W. (2013). Local and regional development in global value chains, production networks and innovation networks: A comparative review and the challenges for future research. *European Planning Studies, 21*(7), 967–988.

Ponte, S., & Sturgeon, T. (2014). Explaining governance in global value chains: A modular theory-building effort. *Review of International Political Economy, 21*(1), 195–223.

Popkin, G. (2019). How scientists can team up with big tech. *Nature, 565*(7737), 665–668.

Pozsar, Z. (2018). Repatriation, the echo-taper and the €/$ basis. *Global Money Notes, 11.*

Rikap, C. (2018). Innovation as economic power in global value chains. *Revue d'économie Industrielle, 163*, 35–75.

Rikap, C. (2019). Asymmetric power of the core: Technological cooperation and technological competition in the transnational innovation networks of big pharma. *Review of International Political Economy, 26*(5), 987–1021. https://doi.org/10.1080/09692290.2019.1620309

Rikap, C. (2020). Amazon: A story of accumulation through intellectual rentiership and predation. *Competition & Change.* https://doi.org/10.1177/1024529420932418

Rikap, C., & Flacher, D. (2020). Who collects intellectual rents from knowledge and innovation hubs? Questioning the sustainability of the Singapore model. *Structural Change and Economic Dynamics*.

Rugman, A., & D'Cruz, J. (1997). The theory of the flagship firm. *European Management Journal, 15*(4), 403–412.

Schrape, J.-F. (2018). Open source communities: The sociotechnical institutionalization of collective invention. In *Collectivity and power on the internet* (pp. 57–83). Springer.

Selwyn, B. (2019). Poverty chains and global capitalism. *Competition & Change, 23*(1), 71–97.

Serfati, C. (2008). Financial dimensions of transnational corporations, global value chain and technological innovation. *Journal of Innovation Economics & Management, 2*, 35–61.

Smith, J. (2016). *Imperialism in the twenty-first century: Globalization, super-exploitation, and capitalism's final crisis.* NYU Press.

Stallkamp, M., & Schotter, A. P. (2019). Platforms without borders? The international strategies of digital platform firms. *Global Strategy Journal*.

Steiber, A., & Alänge, S. (2013). A corporate system for continuous innovation: The case of Google Inc. *European Journal of Innovation Management, 16*(2), 243–264.

Strange, R., & Humphrey, J. (2018). What lies between market and hierarchy? Insights from internalization theory and global value chain theory. *Journal of International Business Studies, 50*, 1401–1413.

Sturgeon, T. J. (2009). From commodity chains to value chains: Interdisciplinary theory building in an age of globalization. In J. Bair (Ed.), *Frontiers of commodity chain research* (pp. 110–135). Stanford University Press. http://isapapers.pitt.edu/84/

Veblen, T. (2017). *The theory of the leisure class.* Routledge.

World Intellectual Property Organization. (2019). *WIPO technology trends 2019: Artificial intelligence.* WIPO.

CHAPTER 4

Tech Giants and Artificial Intelligence as a Technological Innovation System

Abstract In this chapter, we conceptualize artificial intelligence as a *technological innovation system* (TIS) and use data on tech giants' publications to analyse their crucial role in developing and shaping the system. A unique characteristic of the AI system is the integrative role of cloud computing, a set of digital services completely dominated by a handful of tech giants. We demonstrate that tech giants are at the forefront both in the development of artificial intelligence core techniques and in its functional and field applications. We point to the crucial role and limits of digital learning in shaping the future of this TIS. Rather than defining the performance of the system in terms of its expansion, we demonstrate that the dominant position of tech giants results in a problematic guidance.

Keywords Artificial intelligence · Technological innovation system · Big tech · Cloud computing

4.1 Introduction

In this chapter, we use data on tech giants' publications to specify their role in artificial intelligence (AI), drawing upon the literature on *technological innovation systems* (TIS). This literature focuses on the formation, workings and evolution of a more or less broadly defined technology

C. Rikap and B.-Å Lundvall, *The Digital Innovation Race*, https://doi.org/10.1007/978-3-030-89443-6_4

regarded as an innovation system (Bergek et al., 2008, 2015; Markard & Truffer, 2008). For the theoretical framing presented here, we take inspiration primarily from Carlsson and Stankiewicz (1991, p. 111), Liu and White (2001) and Bergek et al. (2008).

AI can be defined both as a knowledge field and as a general-purpose technology with wide potential fields of application. While it started as theoretical reflections within psychology and organization theory, its growing relevance and applicability reflects rapid progress in mathematics, in radical organizational innovations and in a number of complementary information and telecommunication technologies such as the internet, mobile communications and supercomputers (Teece, 2018).

This chapter takes a broad system view in order to capture the specific form of interdependence between knowledge creation and knowledge application that characterizes AI. While feedback from users is crucial for innovation performance in all sectors (Lundvall, 2017), in the case of AI, the training of algorithms and the speed of technical progress is dependent on data inflow, reflecting user activities in connection with large-scale applications. AI is primarily applied in contexts where interactive learning between users and producers can be substituted by algorithms' digital learning on the basis of big data.

We briefly introduce the concept of technological innovation system in Sect. 4.2, and we use it to characterize AI in Sect. 4.3. Section 4.4 uses this framing to analyse the role of tech giants as central actors in the AI TIS. We present the main content of selected tech giants' scientific publications as evidence of the centrality of AI and show that there is synergy between the consolidation of their science and technology lead and their diversification into different fields of AI applications. Section 4.5 focuses on the role of cloud computing as the channel connecting AI producers and users and elaborates on tech giants' capacity to limit users' learning processes. Section 4.6 concludes.

4.2   Technology Innovation Systems

The concept TIS derives from Carlsson and Stankiewicz (1991, p. 111) who defined the technological system as "a network of agents interacting in a specific economic/industrial area under a particular institutional infrastructure or set of infrastructures and involved in the generation, diffusion, and utilization of technology". They presented the concept as complementary to the national innovation system (NIS) with the motivation that technologies differ regarding the geographical extension of interaction—some technological systems are regional or national while others are international or even global. They added to the evolutionary perspective, shared with the NIS-literature, "economic competence" as fundamental to explain the capacity of regions and countries to exploit the potential of emerging TIS.

A decade later, Liu and White (2001) took inspiration from their work and applied their own version of the TIS to an analysis of major changes in China's innovation system—comparing the role of organizations and institutions in relation to different innovation activities before and after the major reforms beginning in the second half of the 1970s. They chose to use the TIS concept because earlier attempts to study China as a NIS, such as IDRC (1997), took the western capitalist model as the ideal and norm, disregarding that organizations (like universities and enterprises) operate differently in different social systems. They proposed that one way to overcome this bias was to focus on *activities* rather than organizations and suggested an analytical scheme encompassing five core activities: (1) research (basic, developmental, engineering), (2) implementation (manufacturing), (3) end-use (customers of the product or process outputs), (4) linkage (bringing together complementary knowledge), and (5) education.

A third version of the concept proposes to study TIS by focusing on specific *functions*. Bergek et al. (2008) argue that such an approach is useful to assess the performance of an emerging technology, where high performance refers to wide diffusion and use. Bergek et al. (2008) identified seven functions: knowledge development and diffusion (F1), guidance of the search (F2), experimentation by entrepreneurs (F3), market formation (F4), creation of legitimacy (F5), resource mobilization (F6), and development of positive externalities (F7).

We are sceptical to the immanent functionalist perspective in this approach. Therefore, we prefer Liu and White's (2001) terminology—referring to activities rather than functions. And we also disagree with the implicit assumption that wider spread and use of a specific technology is always for the better. It reflects a narrow engineering perspective in contrast to a societal perspective, where technical change is evaluated on the basis of its contribution to welfare creation and or problem-solving.

But the scheme may nonetheless be useful for organizing empirical descriptions and policy discussions in relation to a specific technology.

Next, we use the sevenfold scheme rebaptizing Functions as Activities (A1-7) to organize the empirical analysis of the emergence, current state and future of AI. We focus on the role of tech giants in shaping and driving the system. We distinguish between the growth in scale and scope of the system, on the one hand, and performance as related to global welfare and problem-solving, on the other hand.

4.3   ARTIFICIAL INTELLIGENCE
AS A TECHNOLOGICAL INNOVATION SYSTEM

Knowledge development and diffusion (A1) is utterly complex in the AI TIS. AI covers a knowledge field at the intersection of multiple scientific disciplines, ranging from computer science, statistical and mathematic models to biology, linguistics, psychology and neuroscience (Nilsson, 2010). As evidenced by related scientific publications and patents, it is a technology whose development and diffusion are characterized by a recent explosive acceleration. From 1960 until early 2018, there were almost 340,000 patent families and over 1.6 million scientific publications. The recent acceleration in the growth in AI patent publications is impressive. On average, patents grew about 8% a year between 2006 and 2011, and 28% a year between 2012 and 2017 (World Intellectual Property Organization, 2019).

The guidance of AI (A2) reflects a sometimes conflictual, sometimes consensual combination of private motives and the interest of the state. The recent acceleration of its development and use went hand in hand with explosive growth in private internet-related activities, such as search machines, gaming, social media and e-commerce. Governments have played a shifting role in supporting and directing research in AI. Early on, governments supported AI research based on the assumption that it would soon result in breakthroughs in terms of wide application. More recently, as documented in Chapter 7, the state in China and the US as well as the European Commission have become major actors in a global race for AI dominance.

This mix of corporate and core states' guidance has resulted in a focus on specific areas within AI. WIPO (2019) identified five AI techniques: machine learning, probabilistic reasoning, fuzzy logic, logic programming and ontology engineering. Among those, research on machine learning stands out. It was the fastest-growing technique (28%), mentioned in 89% of all AI patent fillings until March 31, 2018. Within machine learning, deep learning is the fastest-growing field in terms of patent filings, with a 175% increase between 2013 and 2016.

The guidance towards this sub-discipline reflects the expectations and incentives of major agents within the system. As mentioned in Chapter 2, Cockburn et al. (2018) explain why the development of machine learning, particularly deep learning and neural networks, should be regarded as the "invention of a method of invention". These technologies, where the central element is that algorithms are trained by being fed with big data, have almost unlimited applications. So far, this process of *digital learning* has been applied on a big scale to social networks, social media, online gaming and streaming, e-commerce and e-payment. Military weapon systems and police surveillance techniques are other fields of early application that reflect core states' interests.

According to Bergek et al. (2008), all relevant applications for the chosen technology should be identified. In the case of AI, one must distinguish between applications at two distinct levels, functional and industrial applications. WIPO (2019) identified nine functional applications (Knowledge representation and reasoning, Planning and scheduling, Control methods, Computer vision, Speech processing, Predictive analytics, Distributed AI, Robotics and Natural language processing). Here again, R&D and patenting have been guided (A2) towards specific applications. Computer vision, which includes image recognition, stands out; it is mentioned in 49% of all AI-related patents. Natural language processing (14% of all AI-related patents) and speech processing (13%) come next.

In terms of industrial applications, using patent data WIPO (2019) identified 20 fields where AI techniques or functional applications can be applied. Transport, telecommunications and life and medical sciences were identified as the three top industries of all AI-related patents, concentrating 15, 15 and 12% respectively. The existence of two levels of applicability—the more generic in terms of functional applications and the field-specific applications that build on the former—open niche opportunities for firms that only or mostly focus on applications within a single field or area. In Chapter 2, we referred to State Grid Corporation of China and Siemens as second-degree intellectual monopolies. The former concentrates AI patents within the energy field, and the latter is the first AI patent holder in the life and medical sciences field.

4.3.1   *On the Geography of the AI TIS*

Globally, AI is highly concentrated in the US and China, in this order and with a significant advantage for the US (Castro et al., 2019; UNCTAD, 2019; World Intellectual Property Organization, 2019). A recent report on AI dominance conducted by the Center for Data Innovation showed that the US arrives first in terms of AI talent, research, development and hardware, while China is the world leader in AI adoption and data (Castro et al., 2019). This dominance of either the US or China is true for the location of research efforts and outcomes as well as for entrepreneurship and resource mobilization.

Entrepreneurship (A3) provides evidence of this advantage. According to China Internet Development Report 2021, there are 1454 AI companies in China and 2257 in the US. A previous study by the China Institute for Science and Technology Policy at Tsinghua University (2018) had already acknowledged significant AI entrepreneurship, mostly concentrated in the US and China. By June 2018, the report found 4925 AI enterprises worldwide, 42% from the US and 20% from China (excluding Hong Kong, Macao and Taiwan). The same report showed that the US has nine of the world's top 20 cities by the number of AI enterprises while China has four. However, Beijing led the ranking by afar with 395 AI enterprises followed by San Francisco with 287 and London with 274. The rest of the ranking included cities from Canada, Germany, France and Israel.

The US also stands out when it comes to well-funded AI companies (defined as those that have received at least 1 million USD in combined funding), with 1727 companies. It is followed by the European Union (762) and China (224) (Castro et al., 2019). The US has the greatest number of AI start-ups and the highest levels of private equity and venture capital received by AI start-ups (Castro et al., 2019). Nevertheless, it must also be noticed that since 2010 there has been a slowdown in the US high-tech entrepreneurship rate (Haltiwanger et al., 2014).

Within and beyond entrepreneurship, substantial resources (A6) are mobilized to develop AI. This reflects high expectations of the associated economic benefits. Predictions estimate that AI could generate an additional USD 13 trillion globally (around 1.2% additional annual GDP growth) by 2030. However, gains associated with the adoption of AI will be uneven and the wider use may lead to growing income inequality at all levels (between regions, firms and households). The US accounts for 66%

of AI private funding (venture capital, private equity and M&A) while China follows from afar but growing fast (it represented 17% of total AI private funding in 2016) (International Telecommunication Union, 2018). By 2019, the US and China concentrated 90% of the market capitalization value of the 70 world's biggest digital platforms (UNCTAD, 2019).

As we can see, currently most of the activities in the AI TIS are dominated by either the US or China. While the AI TIS is global when it comes to its use and market expansion (A4), three of the core activities (knowledge creation [A1], entrepreneurial activities [A3] and resource mobilization [A6]) are concentrated in the US and China. This has important implications for the guidance and direction of the technology (A2) and for its legitimacy (A5) in the rest of the world.

4.3.2 Summing Up on the Characteristics of the AI TIS

At a high level of abstraction, the emergence of AI may be seen as a response to capitalism's permanent attempts to reduce its dependence on human labour. At a lower level of abstraction, it appears to be a supply-driven technology. Market creation for AI use (A4) is driven mainly by firms, and this reflects that there are serious issues related to the acceptability of the technology (A5). Concerns on its legitimacy range from the future of work (Autor, 2015) to governments and private companies' surveillance (Zuboff, 2015, 2019).

Resistance to applying AI is modest in activities where the level of human and material costs connected to the worst outcome is low. Getting slightly wrong when using a search machine or getting wrong friend recommendations on Facebook is tolerable. Acceptability is lower when there is a risk for major costs. While decision-making may be left in the hands of algorithms when it comes to playing chess and ranking options in an e-commerce marketplace according to different consumer segments, it is more problematic to do so when it comes to make diagnoses of life-threatening illness and manage nuclear power plants.

The AI TIS is characterized by positive externalities leading simultaneously to the globalization of markets and extreme degrees of concentration of capabilities in both geographical and corporate terms. Markets for AI are global while there is a geographical concentration of resource mobilization, knowledge development and entrepreneurship in the US and China.

A major issue is how this geopolitical constellation and the combined corporate and state interest in directing AI will guide future R&D efforts and how it will affect the legitimacy of the technology. For instance, to what degree can the EU, through regulatory activities, influence the direction of technological change, when the US and China dominate R&D in the AI TIS? Will the national origin of the technologies become a major factor in determining the legitimacy of the technology in different parts of the world?

So far, guidance has led to a research focus on machine learning, in particular deep learning, to generic applications in the form of computer vision and natural language processing and to a growing focus on transport, telecommunications and health when it comes to industrial applications. In the next section, we will demonstrate that these orientations reflect tech giants' strategies. So far, they have been the prime agents shaping the direction of the AI TIS.

4.4   AI at the Centre of Tech Giants' Research Efforts

AI is at the core of tech giants' activities. By late 2017, Amazon was the company offering most jobs in AI in the US. Microsoft ranked third.[1] Google continuously innovates in ways to produce new data (such as Gmail, Google Maps and even healthcare apps). These innovative datasets keep feeding its business and lead to larger profits. Likewise, Tencent (2018) states that it invests in AI for all its products as well as new endeavours, a strategy that the company calls "ubiquitous AI". In 2017, the MIT Technology Review ranking of the world's 50 Smartest Companies ranked Tencent 8th and Alibaba 41st.[2]

In this section, we analyse the content of selected tech giants' scientific publications. If AI is a priority for them, then AI techniques and functional applications should stand out in their scientific publications. Table 4.1 presents the 25 most frequent multi-terms for each of these corporations' scientific publications until 2019 included. They were obtained by performing a lexical extraction from the abstracts, keywords and titles (detailed methodology in Chapter 5).

---

[1] https://www.glassdoor.com/research/ai-jobs/.

[2] https://www.technologyreview.com/lists-tr50/what-are-the-50-smartest-companies/.

Table 4.1 Publications' most frequent multi-terms for selected tech giants

Google	Amazon	Microsoft	Tencent	Alibaba
machine learning	machine learning	machine learning	neural network	neural network
neural networks	deep neural networks	speech recognition	convolutional neural networks	recommender system
speech recognition	neural network	data sets	social networks	reinforcement learning
deep learning	genetic algorithm	training data	machine learning	user behavior
deep neural networks	data sets	neural networks	benchmark datasets	deep neural network
language model	cloud computing	video coding	training data	convolutional neural network
acoustic models	natural language	language model	neural machine translation	social networks
approximation algorithms	speech recognition	social networks	image retrieval	data sets
learning algorithms	knowledge graph	search engine	big data	natural language
reinforcement learning	convolutional neural network	data center	topic model	e-commerce platforms
training data	acoustic model	image retrieval	attention mechanism	proposed algorithm
mobile devices	training data	natural language	representation learning	big data
recurrent neural networks	data centers	computer vision	computer vision	search engine
natural language	predictive models	deep neural networks	target domain	attention mechanism
search engines	social media	mobile devices	domain adaptation	Online shopping
computer vision	computer vision	data structures	transfer learning	benchmark datasets
automatic speech recognition	approximation algorithms	web search	feature learning	question answering
efficient algorithms	data streams	convolutional neural networks	reinforcement learning	network based
convolutional neural networks	learning algorithms	learning algorithms	learning approach	display advertising

(continued)

Table 4.1 (continued)

Google	Amazon	Microsoft	Tencent	Alibaba
energy efficiency	sentiment analysis	programming language	learning algorithm	user experience
data mining	object detection	search results	search results	representation learning
voice search	big data	software engineers	face alignment	transfer learning
language processing	topic models	recurrent neural network	face images	short text
computational cost	transfer learning	approximation algorithms	community detection	data analytics
cloud computing	word embeddings	web pages	learning methods	recurrent neural networks

Source Authors' analysis based on Web of Science

4.4.1   Tech Giants Focus on Machine Learning

Machine learning (in particular deep learning and neural networks) and big data multi-terms dominate both in terms of frequency (between 7 and 10 terms on machine learning and its techniques for each selected giant) and positioning in Table 4.1. As we pointed out in Sect. 4.3, this AI technique predominates by far globally. Although coming from different origins, chosen corporations have in common that they daily collect billions of data points and focus on AI as their core technology. In terms of functional applications, their most frequent multi-terms deal with computer vision, natural language processing and speech recognition, which are the most predominant functional applications of the AI TIS (see Sect. 4.3).

By focusing on data harvesting and the development of general machine learning techniques, these companies have introduced specific solutions for most of the AI application fields. Table 4.2 shows for each AI application field—as identified by WIPO (2019)—examples of related initiatives conducted by the selected US and Chinese tech giants. This evidence is in line with our argument of digital intelligence as a driver of tech giants' diversification.

Banking and Finance is the only application field where Alibaba and Tencent have a clear lead in comparison with the US tech giants. They have developed an intertwined financial and non-financial business based

Table 4.2 Tech giants' initiatives for AI application fields. Illustrative examples

AI field Applications	Google	Amazon	Microsoft	Alibaba	Tencent
Banking and finance	Google Pay	Amazon Pay	Microsoft Pay	**Ant Group**	**WeChat Pay,** WeSure, WeShare, Waterdrop, **WeBank**
Cartography	**Google Maps**	UpNext (3D mapping, acquired in 2012)	Land cover mapping	HD maps. Acquired AutoNavi (now known as Amap)	Tencent Maps (HD maps team)
Document management and publishing	Google Analytics (Google Cloud's Big Query), Google Translate	AWS Amazon Textract	**Microsoft 365 (Word),** OneNote	–	Tencent Docs
Networks (social networks, IoT, Smart cities)	Google Assistant, Google Nest, Sidewalk labs	Amazon Echo (Alexa), Amazon Alexa Built-in devices, Amazon Go	Microsoft CityNext, Azure Internet of Things	ET City Brain, TmallGenie	WeChat, Net City
Business (Customer service, e-commerce, Enterprise computing)	Google Cloud, Google Search (ads business), Contact Center AI solution, Google Analytics	**Amazon.com, AWS,** Sizmek (third party ad tech provider)	Azure, Microsoft 365, Edge	Alibaba, Taobao, and Tmall. Ali Cloud	Online shopping feature in WeChat app. Tencent Cloud
Personal devices, computing and HCI	HCI contributes to Search, Gmail, Docs, Maps, Chrome, Android, YouTube	Kindle	**Microsoft 365.** "Human Understanding and Empathy" HUE team	AliGenie	–

(continued)

Table 4.2 (continued)

AI field Applications	Google	Amazon	Microsoft	Alibaba	Tencent
Entertainment	You Tube	Amazon prime video, Amazon music	Xbox (gaming)	Youku Tudou (streaming video), Alibaba pictures, Alibaba music, Albiaba sports	**Online gaming** and Tencent Video
Military	Project Maven. (Former Google Chairman serving as Chair of the National Security Commission on Artificial Intelligence)	Rekognition. Cloud Computing for Defense, AWS Secret Region	Joint Enterprise Defense Infrastructure (JEDI) project. Augmented reality headsets for US Army	—	—
Telecommunications	Google Fiber, Google Meet, Contact Center AI solution. Subsea internet cables	AWS Machine Learning for Telecommunication. Subsea internet cables	Microsoft Intelligent telecommunications. Microsoft Teams and Skype. Azure	Alibaba Cloud: Full-Lifecycle Telco Service Automation, Data-Driven Telco Platform	Digital Guangdong Network Construction, joint venture with Chinese telcos
Computing in government	Google Cloud, Big data for statistics agencies	**AWS**	Microsoft Industry, Azure, **Microsoft 365**	Ali Cloud	Tencent Cloud

AI field Applications	Google	Amazon	Microsoft	Alibaba	Tencent
Industry and manufacturing	Digitization of their production facilities and supply chain. For instance, for Renault	Amazon Robotics	Microsoft Industry	Mobile Robots: TmallGenie Space Ball, Space Interstellar	Light 2.0 Program. Private Equity in UBTech. Tencent AI: Ambient Intelligence
Energy management	DeepMind is used to reduce energy consumption in Google's DataCenters	Collaboration between AWS and AutoGrid. "Climage Pledge Arena" in Seattle	Microsoft Industry: Microsoft for energy	Alibaba–AI Lab for Energy (Khalifa University and Alibaba Cloud research lab)	"AI for FEW" (AI used in food, energy and water)
Life and medical sciences	Verily, DeepMind, Google Health. Calico	Amazon Halo, AWS Health, PillPack and Health Navigator, Alexa can detect a cough/cold	Microsoft HealthVault	AliH ealth	Tencent AI Medical Imaging
Security	Chronicle, an AI-driven solution for the cybersecurity industry	Amazon Guard Duty, Harvest.ai, AWS Automated Reasoning Group	Microsoft SEAL. Homomorphic Encryption (HE)	Alibaba Cloud Security Services. Acquired the cybersecurity start-up Chaitin Tech	Smart Security, Security as a Service (Tencent Cloud)

(continued)

Table 4.2 (continued)

AI field Applications	Google	Amazon	Microsoft	Alibaba	Tencent
Transportation	Waymo	Zoox, Aurora, AWS DeepRacer Championship	Cruise. Microsoft Automotive Industry solutions: to drive a forward-looking quantum research project aimed at reducing traffic congestion	Cooperation with Audi and the joint venture between FAW Group Corporation and Volkswagen in data analysis, internet-vehicle platform building, and intelligent urban transport	–
Education	Google for Education, Google.scholar	Alexa in Education, AWS EdStart	OneNote and Microsoft Teams	"help me answer" app	Tencent Classroom, Tencent Education
Agriculture	Google's secretive "moonshot" laboratory, X, has stepped up plans to modernise food production, discussing plans for precision farming technology with the US government	AWS Smart farm, AWS IoT Greengrass	Land cover mapping, FarmBeats	ET Agricultural Brain initiative	iGrow (smart planting solution). Tencent and Wageningen University & Research (WUR) "Autonomous Greenhouse Challenge"
Arts and humanities (includes AI for music)	Artsandculture.google.com, Magenta (tensorflow)	AWS deep composer	DeepSinger	–	–
Law, social and behavioral sciences	Google patents, Google Analytics and Big Query	–	–	Alibaba's TMIC consumer analytics platform	–

AI field Applications	Google	Amazon	Microsoft	Alibaba	Tencent
Physical sciences and engineering	Google's applied science team which combines computer science with physics and biology in: Quantum Computing, Google Accelerated Science, Climate and Energy, and Scientific Computing Tools	–	–	–	–

Source Authors' compilation

on big data and AI. Together, they account for 94% of the mobile payments market in China. As a comparison, Apple pay had 22 million active users in the US in the first part of 2019, while Ali Pay (Alibaba) had 500 million and WeChat Pay (Tencent) 900 million. Alibaba and Tencent are pioneers in exploiting networks and data from other businesses to produce digital financial intelligence. The latter is used to provide financial services at a meagre cost, including millions of small vendors that do not have credit records (Frost et al., 2019) (see Chapter 6 on the Chinese state's promotion of as well as attempts to regulate its tech giants).

Moreover, Tencent is the world's largest gaming company, a business that heavily relies on China's consumption of online games. By 2015, 46% of all the data created and replicated inside China was entertainment data. WeChat is Tencent's other primary source of big data. All the data processed with AI in China are expected to grow at a compound annual growth rate of 68% until 2025 (Reinsel et al., 2019). In 2020, Tencent engaged in technological cooperation with Huawei. Tencent's GameMatrix cloud game platform uses Huawei's Kunpeng processors. Besides improving Tencent's cloud business, their co-innovation laboratory will also explore collaborations in AI and augmented reality in games.[3]

We may also expect these companies to become world leaders in smart cities (Alibaba), autonomous driving (Baidu) and AI applied to healthcare (Tencent and Alibaba). The Chinese government is actively seeking to develop a network of AI platforms in these fields that are supposed to rely on Chinese giants' infrastructure (Ernst, 2020). Alibaba's smart city cloud platform "ET City Brain" collects and analyses data from cameras, sensors, social media and the Chinese government. It provides predictions in healthcare, urban planning and traffic (CBInsights, 2018b). Until late 2019, City Brain served 22 Chinese cities and Kuala Lumpur. Concerning healthcare, Tencent's initiatives were detailed by its CEO in a WIPO report, including "Tencent AI Medical Imaging" used by more than 100 major hospitals in China for clinical validation with over 90% accuracy (Huateng, 2019).

---

[3] Retrieved from https://www.reuters.com/article/us-tencent-huawei-games-idUSKB N21E0BV last access May 4, 2021.

4.4.2   Tech Giants' Engagement in Generic and Specific Functional Application

Among industry-level AI-based solutions (Table 4.2), the first eight applications correspond to what could be defined as the first wave of digitalization. It is here where there is a greater market dominance of tech giants.

Tech giants have initially focused on global AI-supported products that either require few customizations or whose customization is granted by AI algorithms. In general, digital intelligence powers their products so that customization is automatically granted. This is the case of Facebook feeds, of Amazon's and Alibaba's e-commerce platforms, Google search engine, etc. One specificity of Tencent is that, unlike the other analysed tech giants, two of its most frequent multi-terms refer to facial recognition. Tencent is a frontrunner in facial recognition patenting in China and provides this service to China Unicom and WeBank (partly owned by Tencent) (CBInsights, 2018b). Facial recognition is an AI-powered service, where the algorithms take care of customization as they learn from the specific data sources they are trained with.

In other cases, like smart speakers, customization lies in the hands of outsourced companies, such as app developers. Tech giants' most popular AI functional applications are indispensable for smart speakers. By mid-2016 Amazon was almost the single market player, with 93.5% market share. Nowadays, after other tech giants entered this market, Amazon still leads with almost 30% share globally (and above 50% share in the US). By 2019, Google was the second player with almost a quarter of the market, followed by Baidu and Alibaba, with around 10% each.[4] In the case of smartphone voice assistants, Apple's Siri and Google Assistant are the absolute leaders.

Ernst's (2020, p. 20) interviewed managers stated that Alibaba and Tencent had been engaging in the development of AI algorithms and chips, in part driven by the US–China conflict. In 2019, Alibaba launched Ouroboros, an AI chip specially designed for speech synthesis. It is applied to its cloud business and its performance matches WaveNet model in

---

[4] https://www.statista.com/statistics/792604/worldwide-smart-speaker-market-share/ and https://voicebot.ai/2020/04/28/amazon-smart-speaker-market-share-falls-to-53-in-2019-with-google-the-biggest-beneficiary-rising-to-31-sonos-also-moves-up/.

Google datacentres (Demler, 2019). Moreover, Alibaba (2019, pp. 84 and 85) claims that the company is:

> one of the few companies in the world with a proprietary, distributed deep learning platform that has access to consumer insights across diverse businesses involving a rich variety of consumer experiences. As a result, we believe we are in a unique position to develop large-scale commercial use of artificial intelligence, or AI. We have applied various AI technologies across our digital economy to enhance the consumer experience. These enhancements include personalized search results and shopping recommendations empowered by deep learning and data analytics (...) In addition, our AI capabilities enable us to introduce innovative products, such as Tmall Genie, our AI-powered smart speaker.

Overall, our results show that AI innovations are selected tech giants' main science and technology priority and that these companies are both focusing on the most prominent AI techniques (machine learning and in particular deep learning and neural networks) and engaging in the development of the most promising AI functional applications. The fact that selected techs giants privilege those AI techniques with the potential to become a general-purpose method of invention, reflects their aim to become data-driven intellectual monopolies (or IM of 4th degree, see Chapter 2).

4.4.3   Tech Giants' Dependence on Data Harvesting

The dominant role of tech giants reflects that they have been successful in applying AI, among others, to search, social media, entertainment and e-commerce. In these areas, the constant flows of data from consumers and customers feed knowledge development. Tech giants' dominant position in producing digital intelligence depends, so far, on these flows of data. Nevertheless, as regulations move forward in limiting data harvesting, they are also attempting to overcome their reliance on big data.

Transfer learning is one example of how tech giants could overcome their data dependence. The term appears among the most frequent multi-terms of Amazon's publications and of the two Chinese giants. This is a technique where the algorithms use what they have learned from a data source not only to make better predictions in relation to that data's underlying problem but also to transfer that learning to other cases. For instance, large language models (with a great amount of data) are used to deliver higher quality translations for languages with limited amounts of training data. Transfer learning might become a more frequently used technique as regulations limit tech giants' free data harvesting. Among others, Amazon could use its e-commerce AI models to infer market behaviours in regions where it is not yet active. Overall, it broadens the scope of tech giants' Internet of Things (IoT) businesses since AI models trained with data from one company could be transferred to other companies within the same industry.

Beyond transfer learning, AI models with more parameters require fewer data to be trained. The cost of training AI models increases with the model's number of parameters. Tech giants have the lead, in particular Google through DeepMind and Microsoft through OpenAI. Created by Elon Musk, OpenAI is currently backed by Microsoft in a joint project to build supercomputer AI technologies for Microsoft's cloud business.[5] Since this investment, Microsoft gained an exclusive license to OpenAI's GPT-3, a frontier deep learning AI model that produces human-like text, and a group of AI researchers left due to internal tensions over the direction and priorities of OpenAI since Microsoft stepped in. These effects of OpenAI joining Microsoft's corporate innovation system are illustrative of the dynamics presented in Chapter 3.[6]

According to the State of AI latest report, GPT-3 is the most advanced model, it has 175 billion parameters. It costs around USD 1 per 1000 parameters; thus, around USD 10 million to train an AI model with this technology (Benaich & Hogarth, 2020). This favours tech giants not only because they own AI models with greater number of parameters but also because they pre-train the AI models and sell them as services in their clouds.

Additionally, tech giants are leading the developments of federated learning (Benaich & Hogarth, 2020; Jacobides et al., 2021); a deep learning approach where the algorithms learn while processing data that do not need to be hosted together at the same location. Federated learning moves attention from data centralization towards data access since only the latter will eventually be required to train AI models. This

---

[5] https://thenextweb.com/artificial-intelligence/2019/07/23/openai-microsoft-azure-ai/.

[6] https://www.geekwire.com/2020/openai-renamed-closedai-reaction-microsofts-exclusive-license-openais-gpt-3/ and https://www.ft.com/content/8de92f3a-228e-4bb8-961f-96f2dce70ebb.

reduces the costs of transferring data, is safer and algorithms (both internally used and those sold as a cloud service) still become increasingly better.

In all these new developments, tech giants are incumbents owning—and keeping secret—the most breakthrough AI models. Although this advantage was built out of harvesting and centralizing data from individuals and other organizations, in the future, dependence on (centralized) data will be reduced and it seems that those with the most advanced AI algorithms (which are also those with the greatest processing power) will reassure their lead.

4.5   The Actors of the AI TIS and the Unique Role of Cloud Computing

As we mentioned in Sect. 4.2, actors and their network relations are key components of a TIS. In the case of the AI TIS, for reasons explained in previous chapters and as we showed here, *tech giants* have established themselves as central and dominant actors. Leading *research institutions* in AI collaborate with them and many of the most productive researchers find employment with them (Chapter 3). *Most new entrepreneurial firms* specializing in AI aspire to be acquired by a tech giant or to become integrated into their corporate innovation systems. Some of them might nonetheless become *aspiring competitors* as is the case of Bytedance (the owner of TikTok). Finally, *suppliers* of advanced technologies—semiconductors, communication equipment, etc.—required to advance the development and use of AI have an important role in the AI TIS. All these actors interact in producing new AI knowledge.

But this set of actors covers only the supply side of the AI TIS. The other side has to do with the application of AI and the role of users. A key insight from innovation studies is that it is insufficient to focus on the supply side and that interaction with users is central to the rate and direction of innovation (Lundvall, 1985; Rosenberg, 1982; von Hippel, 1988). In the specific case of AI, users' data are crucial for the process of knowledge creation and tech giants' privileged position in harvesting data on a big scale is essential to their overall dominance. There is, thus, an extremely close interconnection between knowledge creation and knowledge application in the AI TIS.

Therefore, *consumers* and other *customer firms* using *free* (paid with data) or paid-for services are essential parts of the AI TIS. This gives another reason why—besides contributing to its advertising business—many Google products are free for users. They provide Google with constant streams of big data. And the same can be said of other tech giants' platforms, like those belonging to Facebook and Tencent. Harvested big datasets are used to further develop their capabilities in the very core of AI technology. As we explain next, another unique phenomenon characterizing the AI TIS is cloud computing. It takes on a crucial role to mediate digitalized interaction between professional AI users and AI suppliers.

4.5.1   Cloud Computing and the AI TIS

The public cloud represents a unique form of network interconnecting different agents within the AI TIS, facilitating the development of field applications. The cloud works as a channel from tech giants' generic AI to functional applications and required adaptations to local cultural and institutional contexts. This results in an interconnected AI TIS where companies developing specific AI applications are dependent on tech giants' AI-generic solutions. As the AI TIS further develops such patterns, tech giants' dominance becomes even stronger. An extreme example demonstrating the political power of the firms hosting the cloud was the recent case of Parler, the social network popular among right-wing supporters in the US. After the US Capitol attack, AWS stopped providing cloud services to Parler, which basically meant shutting this social network down.

Similarly, Netflix (2019, p. 10) stated in its 2019 annual report that its software and computing systems depended on data processing, storage capabilities and other services provided by Amazon Web Services (AWS) and that it could not easily switch to another cloud provider. Altogether, this dependence represented a potential risk for its business, probably all the more risky considering that Amazon is its competitor in the video streaming market.

As we introduced in Chapter 1, the public cloud refers to software, hardware and platforms that private third-party providers offer as online services. Only a handful of tech giants dominate the global market for these kinds of digital services. AWS has around 34% of the public cloud market, followed by Microsoft. Amazon's profits are mostly coming from this service, while most of its revenues come from its e-commerce activity (Amazon, 2020). AWS has Facebook, Twitter, Adobe, Netflix, Spotify and Zoom as customers. While Amazon's market share remained steady in recent years, Microsoft's went from less than 10% at the beginning of 2016 to almost 20% by the end of 2019. Alibaba is the fourth player with more than 5% of the market, preceded by Google (less than 10%). The overall result is an extreme and growing concentration of this global market around these four players. By the end of 2020, these four companies represented 70% of the public cloud computing world market, up from 57% at the beginning of 2016 (Synergy Research Group, 2019, 2021).

In the public cloud, clients pay according to what they use. Clients save tangible capital investment, while tech giants enjoy economies of scale and scope. Since the reproduction costs of those services tend to zero, as these companies expand their client base, profits increase exponentially. The cloud computing market potential is huge: CBInsights (2018a) estimates that by 2022 cloud computing will be a 513 billion USD market and cloud storage 90 billion USD. Public cloud computing represents around 80% of the total cloud infrastructure, growing 40% annually (Synergy Research Group, 2019).

Tech giants centralize AI provision through their cloud businesses (Jacobides et al., 2021). As a rented service, the public cloud offers customer access to graphic processing units (GPU) that are crucial for the application of AI. For those companies that do not have the internal capacities or prefer to outsource the training process, tech giants also offer pre-configured virtual machine templates (that run on the GPUs) to train AI algorithms. They offer "data as a service" by providing access to standardized databases (for instance image datasets) to train algorithms, which are essential for building AI applications. And they offer data preparation tools that set the data in a format suitable for being processed by the algorithms. Moreover, tech giants offer growing sets of AI services in their clouds that assist AI applications' development. Summing up, AI as a service in the cloud is central for allowing a myriad of companies to apply the AI core developed by tech giants to specific AI applications. Since technologies are lent as services and code is generally not open access, the learning by using and reverse engineering chances of those developing AI applications and more generally using cloud services are reduced.

At the same time, the cloud computing business is a means that tech giants can use to identify new promising businesses. From time to time, tech giants copy services developed by third companies that become preferred by cloud users. This was the case of Elastic's Elasticsearch and

Kibana services offered directly by AWS.[7] Moreover, cloud computing allows tech giants to early notice when companies using the services are thriving. For this purpose, they can use the information on storage and processing power for data and on the use of different AI services. The cloud offers them an open door to sneak into (and copycat) thousands of organizations around the world.

4.6   Concluding Remarks

In the first part of the chapter, we presented AI as a TIS. First, we showed that the system is expanding with great speed both in size and scope. Second, we pointed to some unique characteristics of this TIS when it comes to the synergy between technology development and market formation. Platform organization and the public cloud result in positive externalities giving early movers and incumbents dominant positions.

As a result, the US and Chinese tech giants have become the "articulators" and even planners of the system. They play crucial roles in most of the key activities of the AI TIS. They exploit positive externalities (A7) and they dominate development and diffusion (A1), market formation (A4) and resource mobilization (A6). Through their corporate innovation systems, they draw upon external entrepreneurial efforts (A3). As a result, they have a prominent role in guiding search and the direction of new applications (A2). Their main weak point relates to their legitimacy (A5) reflecting growing public resistance to their use and abuse of data.

Therefore, it is significant that tech giants are making efforts to use their unique capabilities to become less dependent on big data, simultaneously as they diversify the fields of applications and extend the use of AI (and data harvesting) to other sectors such as health, education, transport, energy, environment and urban management. A crucial issue for the scale and speed of this extension is, besides legitimacy issues, to what degree digital learning can be applied in these new areas and substitute for forms of interactive learning between technology producers and users.

For each of the tech giants, their current focus on AI core or most generic techniques (Google calls it a focus on "artificial general intelligence") and functional applications reflects a unique evolutionary

---

[7] Retrieved from https://www.theregister.com/2021/01/22/aws_elastic_fork/ last access May 20, 2021.

capacity-building process. In the next chapter, we elaborate on the technological convergence of Microsoft and Amazon within the AI TIS. Amazon and Microsoft, with a quite distinct original business focus, saw the potential in organizing research and promoting innovation in AI and cloud services. Their convergence illustrates how tech giants move along a technological trajectory at the core of the evolution of the AI TIS.

REFERENCES

Amazon. (2020). *Amazon Annual Report 2019*. Amazon.

Autor, D. H. (2015). Why are there still so many jobs? The history and future of workplace automation. *Journal of Economic Perspectives, 29*(3), 3–30. https://doi.org/10.1257/jep.29.3.3

Benaich, N., & Hogarth, I. (2020). *State of AI Report*.

Bergek, A., Hekkert, M., Jacobsson, S., Markard, J., Sandén, B., & Truffer, B. (2015). Technological innovation systems in contexts: Conceptualizing contextual structures and interaction dynamics. *Environmental Innovation and Societal Transitions, 16,* 51–64.

Bergek, A., Jacobsson, S., Carlsson, B., Lindmark, S., & Rickne, A. (2008). Analyzing the functional dynamics of technological innovation systems: A scheme of analysis. *Research Policy, 37*(3), 407–429.

Carlsson, B., & Stankiewicz, R. (1991). On the nature, function and composition of technological systems. *Journal of Evolutionary Economics, 1*(2), 93–118.

Castro, D., McLaughlin, M., & Chivot, E. (2019). Who is winning the AI race: China, the EU or the United States? *Center for Data Innovation*.

CBInsights. (2018a). *Amazon strategy teardown*.

CBInsights. (2018b, April). *Rise of China's big tech in AI: What Baidu, Alibaba, and Tencent are working on*.

China Institute for Science and Technology Policy at Tsinghua University. (2018). *China AI development report*.

Cockburn, I. M., Henderson, R., & Stern, S. (2018). *The impact of artificial intelligence on innovation*. National Bureau of Economic Research.

Demler, M. (2019). Alibaba's Ouroboros is a fast talker. *Linley Newsletter*. www.linleygroup.com/newsletters/newsletter_detail.php?num=6058&year=2019&tag=3

Ernst, D. (2020). *Competing in artificial intelligence chips: China's challenge amid technology war*. Centre for International Governance Innovation.

Frost, J., Gambacorta, L., Huang, Y., Shin, H. S., & Zbinden, P. (2019). *BigTech and the changing structure of financial intermediation*. Bank of International Settlements.

Haltiwanger, J., Hathaway, I., & Miranda, J. (2014). Declining business dynamism in the US high-technology sector. *Available at SSRN 2397310*.

Huateng, M. (2019). Application of artificial intelligence and big data in China's healthcare services. In S. Dutta, B. Lanvin, & S. Wunsch-Vincent (Eds.), *GLOBAL INNOVATION INDEX 2019: Creating healthy lives—The future of medical innovation* (pp. 103–110). WIPO.

International Telecommunication Union. (2018). *Assessing the economic impact of artificial intelligence* (ITU Trends. Emerging Trends in ICTs.). International Telecommunication Union.

Jacobides, M. G., Brusoni, S., & Candelon, F. (2021). The evolutionary dynamics of the artificial intelligence ecosystem. *Strategy Science*.

Liu, X., & White, S. (2001). Comparing innovation systems: A framework and application to China's transitional context. *Research Policy, 30*(7), 1091–1114.

Lundvall, B. A. (1985). *Product innovation and user-producer interaction*. Aalborg University Press. http://vbn.aau.dk/ws/files/7556474/user-producer.pdf

Lundvall, B. A. (2017). *The learning economy and the economics of hope*. Anthem Press.

Markard, J., & Truffer, B. (2008). Technological innovation systems and the multi-level perspective: Towards an integrated framework. *Research Policy, 37*(4), 596–615.

Netflix. (2019). *Annual Report*.

Nilsson, N. J. (2010). *The Quest for artificial intelligence: A history of ideas and achievements*. Cambridge University Press.

Reinsel, D., Wu, L., Gantz, J., & Rydning, J. (2019). *The China datasphere: Primed to be the largest datasphere by 2025* [White paper]. IDC.

Rosenberg, N. (1982). *Inside the black box: Technology and economics*. Cambridge University Press.

Synergy Research Group. (2019). *Amazon, Microsoft, Google and Alibaba strengthen their grip on the public cloud market*. https://www.srgresearch.com/articles/amazon-microsoft-google-and-alibaba-strengthen-their-grip-public-cloud-market

Synergy Research Group. (2021). *Cloud market ends 2020 on a high while Microsoft continues to gain ground on Amazon*. Synergy Research Group.

Teece, D. J. (2018). Profiting from innovation in the digital economy: Enabling technologies, standards, and licensing models in the wireless world. *Research Policy, 47*(8), 1367–1387.

Tencent. (2018). *2017 Annual Report: Smart communication inspires*.

UNCTAD. (2019). *Digital Economy Report 2019: Value creation and capture—Implications for developing countries*. United Nations.

von Hippel, E. (1988). *The sources of innovation*. Oxford University Press.

World Intellectual Property Organization. (2019). *WIPO technology trends 2019: Artificial intelligence*. WIPO.

Zuboff, S. (2015). Big other: Surveillance capitalism and the prospects of an information civilization. *Journal of Information Technology, 30*(1), 75–89.

Zuboff, S. (2019). *The age of surveillance capitalism: The fight for a human future at the new frontier of power: Barack Obama's Books of 2019*. Profile Books.

Amazon and Microsoft: Convergence and the Emerging AI Technology Trajectory

Abstract In Chapter 4, we analysed artificial intelligence (AI) as a technological innovation system (TIS) dominated by the tech giants. This chapter gives insights into the emergence and dynamics of this system. We explore the technological convergence between two tech giants with quite distinct origins using lexical analyses of these companies' patents and scientific publications. We find that both Amazon and Microsoft have zoomed in their research and development (R&D) efforts on deep learning and neural networks as well as functional AI applications. We also find evidence of increasing centrality of harvesting, storing and processing data. R&D on cloud computing infrastructure is another area where both companies overlap. Given their dominant role in the AI TIS and the importance of economic factors in the selection of the cluster of technologies that constitute technological paradigms, we argue that these companies' priorities are indicative of the prevailing directions within AI technological trajectories.

Keywords Data-driven intellectual monopoly · Rent-extraction · Corporate innovation system · Amazon · Microsoft

© The Author(s), under exclusive license to Springer Nature Switzerland AG 2021
C. Rikap and B.-Å Lundvall, *The Digital Innovation Race*,
https://doi.org/10.1007/978-3-030-89443-6_5

5.1   INTRODUCTION

Microsoft and Amazon have established themselves as world leaders in artificial intelligence (AI) and they have conquered quasi-monopolistic positions in the market for cloud computing. To compare the technological paths that they have followed is a way to track the technological evolution of the AI technological innovation system and the most recent tendencies give interesting insights into the directions of the technological trajectories. As we shall see, Microsoft's and Amazon's technological profiles have become quite similar despite their distinct industrial origin— a software monopolist vs an e-commerce book shop. This reflects the strength of convergence of AI-related technologies forming a new technological trajectory within the ICT technological paradigm distinct from the first one, which was dominated by micro-electronics.

5.1.1   *Technological Paradigms and Technological Trajectories*

We take inspiration from Giovanni Dosi's (1982) concept of *technological paradigms*, introduced as an analogy with Kuhn's scientific paradigms and defined as "an 'outlook' which defines the relevant problems, a 'model' and a 'pattern' of inquiry". Once the new technological paradigm is established, Dosi (1982) claims that incremental innovations are identified as the normal technical progress, as opposed to radical innovations that characterize the emerging technological paradigms. He continues arguing that technological paradigms are characterized by a stage where new clusters of technologies emerge, competing with the previously dominant one and economic factors become determinant for the selection of the cluster of technologies that will constitute the new paradigm.

Dosi (1982) also introduced the concept of *technological trajectory* as the established direction of technological change characterizing a technological paradigm. His conceptualization of the innovation process had roots in his own empirical work on semiconductor technologies where he found that technical progress could be defined in terms of improvement of the trade-off between density and computing power (Dosi, 1981). In what follows, we find a similar focus on improving the trade-off between data volume and machine learning.

Dosi (1982) links shifts in technological paradigms to changes in the industrial structure where the formation of a new paradigm is characterized by waves of entrepreneurial activities while the stabilization results in oligopolistic structures. Judging from this perspective, the dominance of data-driven intellectual monopolies indicates that the current AI paradigm is already mature. An alternative interpretation is that this paradigm is different with tech giants setting the direction of AI's overall trajectory and profiting the most from its results.

5.1.2 Two Tech Giants' Technological Convergence

Microsoft was a frontrunner of the ICT revolution's first phase. Its intellectual monopoly historically relied on patenting and copyrights (Arthur, 2012). For decades, high margins depended on exclusivity over Windows and the Office package, relying on routine innovations (Pisano, 2015). However, the move from desktop to mobile handsets left Microsoft behind Google and Apple (Arthur, 2012). Furthermore, Microsoft's dominance was challenged by open-source software (OSS) and by piracy. 2014 was a turning point as the company moved to become a data-driven intellectual monopoly. Microsoft deprioritized Windows and gave priority to mobile and cloud businesses based on big data and AI (in particular deep learning) as sources of continuous innovations (Ibarra & Rattan, 2018).

In the meantime, Amazon emerged as the absolute e-commerce leader. Amazon's data advantage comes from its globally dominant e-commerce business, concentrating more than a third of global online retail (UNCTAD, 2019). Since 2002, Amazon uses automated systems based on processed and analysed market data to provide recommendations to consumers (World Intellectual Property Organization, 2019). Amazon can match consumer habits with personal information, and it has been claimed that it discriminates prices using that data (Khan, 2017). Supply-side network effects favour Amazon's extraction of value from unequally exchanging with third-party sellers. These companies depend on Amazon for market access; they are under pressure to accept Amazon's fees, sales and payment conditions (Rikap, 2020). E-commerce is Amazon's main source of data, while—as we mentioned in Chapter 4—its main source of profits is Amazon Web Services (AWS), its cloud business. Although Amazon and Microsoft have become rivals in cloud computing, the former has a clear market leader position (see Chapter 4).

In this chapter, we focus on how these two giants have moved from their originally distant industrial fields to focus on the same key elements in AI and digital services. This illustrates what Srnicek (2017) has called the convergence thesis. According to the author, the need for more data gives direction to platforms' expansion resulting in a "tendency for different platform companies to become increasingly similar as they encroach upon the same market and data areas" (Srnicek, 2017, p. 59). Cloud computing is an apparent example of this convergence. Nonetheless, our convergence thesis goes beyond Srnicek's (2017) idea of market and data convergence in two respects. First, not only markets and data, but also tech giants' technological trajectories are converging. Second, we consider this technological convergence as taking place only in those platform companies that have secured a vantage position in terms of AI models, processing power and data variety and centralization.

As we showed in previous chapters, tech giants achieved their position through increased tangible and intangible capital investments. Hereby they enjoy vast computational power in centralized infrastructures. They hired experts developing AI machine learning models kept in secret, monetized knowledge produced at the level of their corporate innovation systems and harvested massive amounts of diversified data. All the latter required colossal available funds, which is why forerunners not only have a technological advantage[1] but also an economic advantage. Furthermore, not every source of big data is equally monetizable and neither does it result in the same type of digital intelligence. E-commerce and search engines provide greatly diverse and multi-purpose data, while ride-hailing or vacation rental marketplaces variety of data is limited even if they may provide sizeable data sources.

We go beyond Srnicek's (2017) perspective in another respect. We see the technological convergence between tech giants as reflecting a formation of new technological trajectories within the AI technological innovation system. Technology development efforts are increasingly concentrated in data analysis, cloud computing and specific functional and industrial applications.

Considering these characteristics, we provide evidence of tech giants' convergence by analysing the underlying evolution of Amazon and Microsoft's science and technology priorities. While convergence intensifies technological competition among the tech giants, it does not reduce

[1] This advantage, as we explained in previous chapters, is explained by knowledge cumulativeness and, in particular for the digital sector, by network effects that results in more harvested data and therefore "more productive refineries"—more accurate AI models.

their market power. On the basis of unique access to data and key technologies, they remain intellectual monopolies.

For instance, Microsoft's acquisition of LinkedIn, besides providing a prominent patent portfolio (542 patents), has provided a source of big data that reinforces Microsoft's intellectual monopoly and rents. Through LinkedIn, Microsoft enjoys curriculum data of 690 million people and profiles from 50 million companies. LinkedIn chief executive claimed that five million new data points are entered per minute. Using these data, Microsoft found opportunities to offer to its LinkedIn users training on required skills for 15 USD per certificate, an offer expected to be taken up by 25 million people.[2]

Every Amazon business relies on data and is a source of data, from Kindle[3] to Amazon's home appliances (such as Alexa) and wearables. They are not only a business in themselves but also new sources of data. By using people's voices, Alexa currently controls home appliances, cars, information seeking, entertainment and ordering from restaurants and shops. Concerning wearables, Amazon launched Halo in 2020, a wellness tracker that not only takes 3D scans and monitor's sleep; it is even capable of monitoring our mood.

Beyond specific examples, data-driven intellectual monopolies are permanently on the outlook for data sources. Data are users' means of exchange in the digital economy, while they are a privileged means of accumulation for the tech giants.

To account for Amazon and Microsoft's technological convergence, we analyse their scientific publications and patenting developments over time. We retrieved these corpora from the Web of Science and Derwent Innovation databases and analysed them using network and lexical analyses techniques.

The rest of the chapter is organized as follows. Next, we present our methodology. Sections 5.3 and 5.4 study the content of Microsoft and Amazon's publications and patents, respectively. Finally, Sect. 5.5 concludes on their technological convergence.

---

[2] https://www.ft.com/content/5dbd70e0-c847-4b36-a2fe-2cba5294a7c7.

[3] https://www.theguardian.com/technology/2020/feb/03/amazon-kindle-data-reading-tracking-privacy.

5.2   METHODOLOGY

In order to study Microsoft and Amazon's technological convergence we analysed the content of their codified knowledge. We retrieved information on every scientific publication authored by each of the chosen corporations or any of its subsidiaries until 2019 included from the *Web of Science* and their patents from *Derwent Innovation*.[4] Retrieved patents correspond to every patent from each company's respective first year with data until 2017 included for Microsoft and April 2018 included for Amazon. Both granted and applied patents were retrieved since our interest is their prioritized R&D regardless of whether they apply for patents or not. A summary of the retrieved corpora is presented in Table 5.1.

All the data were processed with the CorText platform, which is an integrated application dedicated to the processing and treatment of various types of corpora.[5] We used it to perform a lexical analysis on the content of each corporation's (1) patent portfolio, and (2) scientific publications' corpora, applying text mining techniques. Van Looy and Magerman (2019, p. 930) define text mining as "the automated extraction of information from text in order to reveal patterns that are present but not obvious in a document collection".

Table 5.1 Summary of patents and publications' corpuses

	Publications	*Patents*
Amazon	824	10,063
Microsoft	17,405	76,109

Source Web of Science and Derwent Innovation

---

[4] Our access to this database included the following patent offices: USPTO, WIPO, European, Japan, Australian, British, Canadian, French, German, Russian and Korean patent offices.

[5] CorText is an open platform for performing bibliometric and semantic analysis that uses the spatial algorithms that draw on classic graph visualization methods for depicting the network maps (Fruchterman–Reingold). It can be accessed online at https://www.cortext.net/.

5.2.1 Patents and Publications' Content Analyses

When analysing patents with mining techniques, Madani and Weber (2016) show that citation analysis has been the most frequently used methodology. The authors explain that the analysis of the actual context of the patent ("patent mining") through techniques like lexical or corpus approaches is a new phenomenon. Furthermore, they argue that lexical analyses are more accurate techniques capable of extracting keywords from analysed text and of recognizing semantic patterns. It is because of these advantages that patents' lexical analysis has been used for technology trend identification (Choi et al., 2011; Yoon et al., 2011).

For each company, we performed a lexical analysis of the titles and abstracts of its patent portfolio. In the case of scientific publications, each lexical analysis considered titles, abstracts and keywords. For each corpus and company, we extracted the top 1000 multi-terms of up to five words restricted to terms that appeared at least once in three different patents/publications. Monograms were excluded and each list was refined following an in-depth cleaning process in order to avoid words not related to the field and whose frequency responds to either their grammatical function ("and", "or", etc.) or the level of grammaticalization within the innovation genre ("improving efficiency", "general purpose", "seamless experience", "use of the application", "other uses", etc.) or the scientific publications' genre ("address these limitations", "article studies", "best known solution", "case study", etc.), respectively. Resulting lists of patents' multi-terms have 811 terms for Microsoft and 856 for Amazon. Publications' multi-terms lists have 757 terms for Microsoft and 648 for Amazon.

For each corpus, network maps were built with those multi-terms that were the most frequently connected among the list of most frequent multi-terms in order to overcome text mining limitations in relation to homonymy, polysemy and synonymy problems and the hyponymy/hypernymy problem.[6] To avoid these issues, we do not look

---

[6] "Words can have different meanings, the same meaning or concept can be expressed by different words, and words can have semantic associations in a hierarchical relation (e.g., animal versus mammal versus cat and dog). Last, not all words are of equal importance in deriving the meaning of phrases, but words with less significance appear with very high-frequency" (Van Looy & Magerman, 2019, p. 932).

at similarities at the document level but at the level of clusters in our network maps.

We plotted selected and refined most frequent multi-terms for each corpus in a network map. In the maps, terms are grouped in clusters. Clustering is a big data analysis technique that builds communities (Fortunato & Hric, 2016). In our case, this meant that they contribute to inferring the meaning of terms that could have multiple meanings and interpret each cluster as a specific technology or topic (Testoni et al., 2021).

To build network maps using patents' most frequently connected multi-terms, each corpus was split into three different periods with the same number of patents in each period. We built network maps for each corpus for each period using the resulting list of 811 multi-terms for Microsoft and 856 multi-terms for Amazon. The network maps depict the most frequently connected multi-terms according to their frequency of co-occurrence. Nodes represent patents' multi-terms and we prioritized the top 100 multi-terms for each period.

To provide a clear sense of topics' evolution within chosen companies' patent portfolios, we included a "tubes layout" (Figs. 5.1 and 5.4) automatically generated by Cortext for dynamic network analysis. It is called tubes because clusters' evolution is traced by connecting related clusters from different periods in a tube format. This layout provides a simplified depiction of clusters' evolution over the three chosen periods for each corporation. Clusters are presented as rectangles, are identified by their most frequent terms and placed in the year corresponding to the midpoint of each period. The tubes layout shows how clusters evolve over time. Two or more clusters may merge into one, one cluster may split into multiple clusters and the same cluster may show a continuation or end at some point in time. The width of the tubes that connect different clusters is proportional to their number of records. The colour of the tubes refers to the degree of similarity between two connected clusters. Darker tubes are more robust in the sense that more nodes are shared between the connected clusters in two consecutive periods.

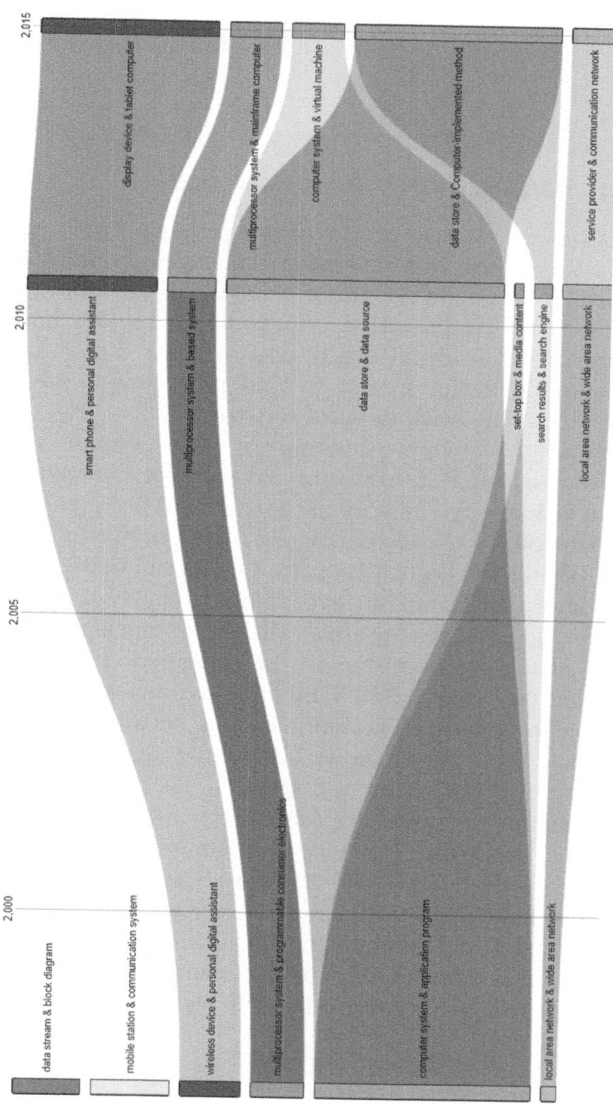

Fig. 5.1 Evolution of Microsoft's patent portfolio content. Tubes layout (*Source* Authors' analysis based on Derwent Innovation)

In turn, for each publications' corpus, we built two multi-terms' network maps. One corresponding to 2014–2017 and the other one to 2018–2019 publications (years for which we do not have detailed patent data).

5.3   Microsoft's Technological Trajectory

5.3.1   *Microsoft's Patent Portfolio Analysis*

Figure 5.1 summarizes the content of Microsoft's patent portfolio. Each rectangle is a cluster of the most frequently connected multi-terms extracted from Microsoft's patents (Figures including all the multi-terms that integrate each cluster are presented in Appendix 5.1). Our interest here is not to provide an in-detail description of every dimension of Microsoft's technological evolution but to find prevailing trends and compare them—in Sect. 5.4—with those of Amazon.

Microsoft's innovation strategy—judging by the content of its patents—remains focused on ICT hardware and software. Nevertheless, its research content changes over time privileging AI and becoming data-driven. Indeed, Microsoft has the second-largest AI patent portfolio, only surpassed by IBM (World Intellectual Property Organization, 2019).

In particular, topics that stand out and gained space over time are related to cloud servicing and data centres. Microsoft's intelligent cloud business is behind all the green clusters in the third period, which are tightly interconnected (see Figure 5.9). The light-green cluster refers to technologies dealing with network communications between the cloud and Microsoft's clients. This cluster shows Microsoft's steady investment in digital infrastructure related to its web services. The content of this cluster evolved as Microsoft moved its business to the cloud. This is observed, for instance, in the disappearance of the multi-term "local area network" in the third period. The multi-term "web service" belongs to this cluster and is also linked to the big green cluster where it was in the second period signalling that this multi-term is a bridge institution[7] in Microsoft's patent content network. This finding reinforces our assertion of cloud computing as one of Microsoft's current priorities.

---

[7] In network analysis, nodes occupying bridging positions are nodes that connect different clusters. They are of particular relevance for holding the clusters together and "in the dynamics of spreading processes across the network" (Fortunato & Hric, 2016).

The other small green cluster of the last period reflects Microsoft research on datacentres infrastructure with multi-terms like "virtual machine" and "computing environment", while the big green cluster refers to Microsoft's software, in particular, software as a service offered through the cloud. Office package for enterprises became a cloud service in 2011. Yet, it was not until 2015 that its penetration jumped to 25%, surpassing Google Apps for business.[8] The conversion of Office into a service increased Microsoft's flow of customer data, transforming how user–producer interaction takes place, in connection with product and service innovation (see Chapter 4). Microsoft increasingly uses customers' data as real-time feedback for incremental innovations. This big green cluster at the bottom of Fig. 5.9 is the largest of the third period. It reflects technologies that interlink cloud services with data management (in the network map, 6 out of the 7 multi-terms include the word "data").

Summing up, patents' lexical analysis shows Microsoft's changing R&D priorities over time, with an increasing focus on the AI TIS. However, not every innovation is patented and there are R&D priorities that cannot be captured by focusing on patents. The following section complements our analysis by studying the content of Microsoft's recent publications.

5.3.2 Microsoft's Recent Scientific Publications

We built a network map with Microsoft's publications 100 multi-terms with the highest frequency of co-occurrence for the period 2014 and 2017 (Fig. 5.2), thus coinciding with the third period of the patent lexical analysis (Fig. 5.1).

Figure 5.2 also depicts Microsoft's most frequent co-authors for each cluster. Google is among the top co-authors in a cluster dealing with more general or basic research on language analysis and processing, confirming that technological cooperation with other data-driven intellectual monopolies is more likely for general knowledge modules. This cluster is connected to two others (blue and red) that reflect Microsoft's research on AI. In line with Chapter 4, multi-terms on neural networks

---

[8] https://www.geekwire.com/2016/five-years-launch-microsofts-office-365-popular-enterprise-cloud-service/.

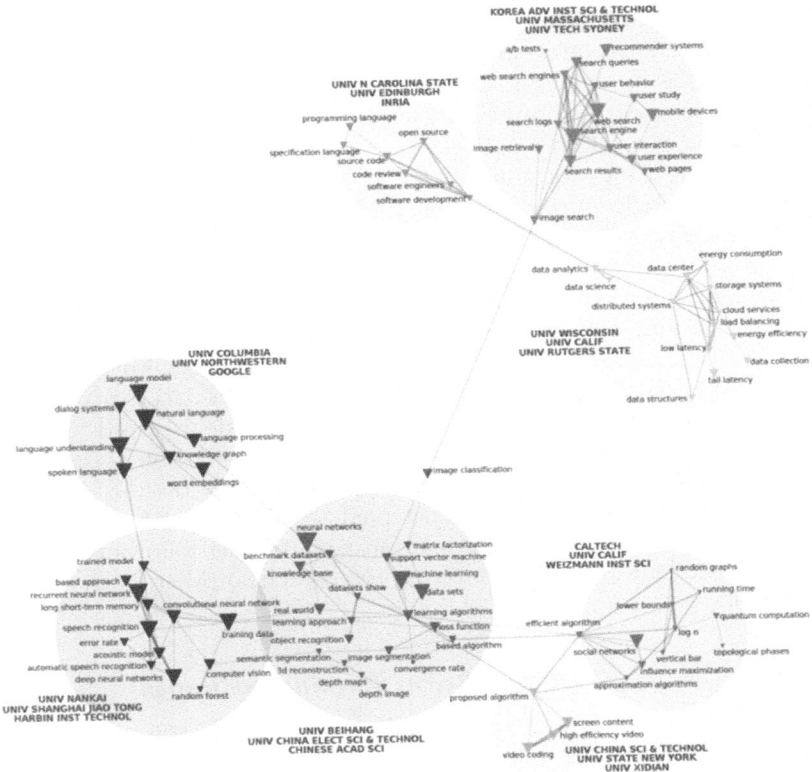

Fig. 5.2 Microsoft's scientific publications (2014–2017) lexical analysis (*Source* Authors' analysis based on Web of Science)

and deep learning stand out both in terms of high frequency and bridging position. Both clusters also include multi-terms with the word "data" and are connected by the multi-term "training data". Microsoft had claimed to have the largest public face recognition dataset. This training data was put down after complaints from individuals who had not given permission to use their faces to train facial recognition algorithms.[9]

There is a third cluster (in yellow) that also includes multi-terms with the word "data". It concentrates most of the "data" multi-terms, from research on data transportation and storage (multi-terms "data center", "storage system" and "energy efficiency") to data analysis ("data analysis" and "data science").

We also built a network map with the 100 multi-terms with the highest frequency of co-occurrence in Microsoft publications between 2018 and 2019 (Fig. 5.3) to capture the most recent trends. Overall, some new

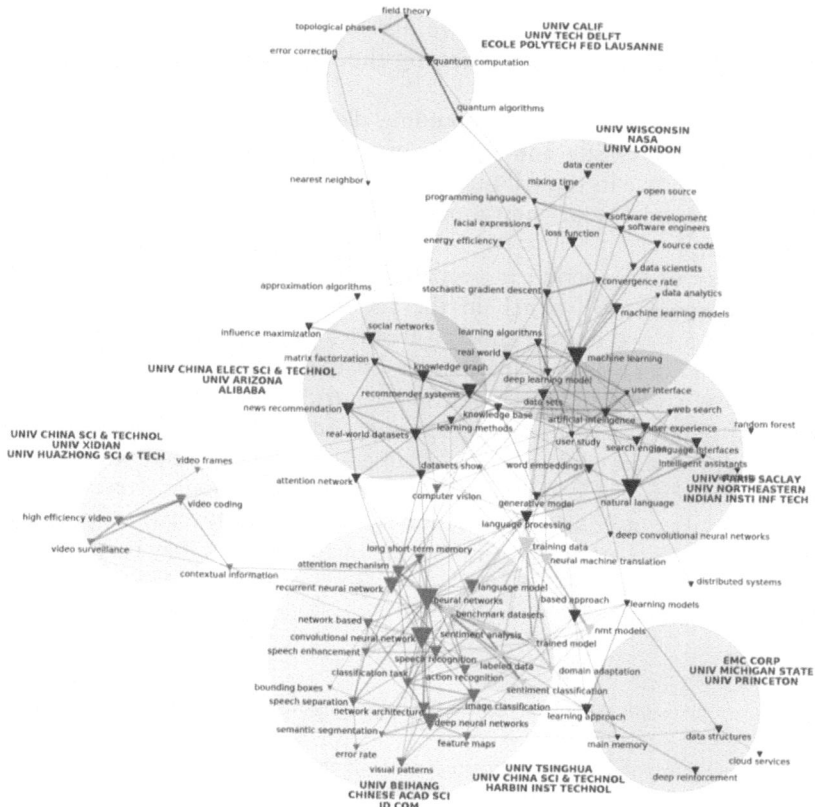

Fig. 5.3 Microsoft's recent publications (2018–2019) lexical analysis (*Source* Authors' analysis based on Web of Science)

trends emerge together with a continuation of Microsoft's already established R&D fields and topics. One difference is that the network becomes more interconnected, pointing to a closer interrelation among Microsoft's published research topics.

Judging by the size of each cluster, in 2018–2019 Microsoft seems to be increasing its efforts in developing technologies related to deep neural networks (big green cluster at the bottom left) *vis-à-vis* cloud servicing basic infrastructure (small blue cluster on the bottom right). AI associated with language and word processing merges with terms related to user experience while the focus on basic or more generic research on "language programming" and "software development" for machine learning remains. "Machine learning" is among the multi-terms with the highest frequency of appearance.

Figure 5.3 also plots a new "training data" cluster (next to the deep neural netwokrs' cluster and in light green). It reflects that Microsoft is seeking to improve its digital intelligence (other multi-terms in the cluster are "trained model" and "nmt models", nmt standing for neural machine translation).

A novelty of this period is a separate cluster for quantum computing (blue cluster on top), which in the previous period was only a topic included in a more diverse cluster (green cluster in Fig. 5.2). Quantum computing is crucial since it provides greater processing power for AI. In this area, Microsoft and Google are frontrunners. Microsoft ranks first in patent filings for quantum computing and has launched its quantum computing services embedded in Azure, offering quantum simulations that run on classical computers.[10] Microsoft and Google's research on this topic takes place in collaboration with leading universities from the US and Europe.[11] Although rivals in the quantum race, they partnered to

---

[10] https://www.infoworld.com/article/3233484/inside-microsofts-quantum-computing-world.html.

[11] Retrieved from https://www.nature.com/articles/d41586-019-03213-z https://www.expresscomputer.in/features/who-holds-the-maximum-quantum-computing-patents-applications/45790/ and https://www.microsoft.com/en-us/quantum/quantum-network last access April 15, 2020.

train developers, so that they can start developing quantum apps.[12] Also, to expand adoption, Azure Quantum includes an OSS toolkit.

As will be shown in Sect. 5.4, the evolution of Microsoft's R&D profile has—with the exception of quantum computing—resulted in a portfolio similar to Amazon's, signalling these companies' technological convergence and indicating that these tech giants both shape and follow a common AI technological trajectory.

5.4   AMAZON'S TECHNOLOGICAL TRAJECTORY

5.4.1   *Amazon's Patent Portfolio Analysis*

Amazon's diversified business is based on an equally diverse innovation portfolio. Figure 5.4 summarizes our lexical analysis of its patent data. The network maps plotting the 100 multi-terms with the highest frequency of co-occurrence for each period are presented in Appendix 5.1 (see Figures 5.10 to 5.12).

In the first period (1996–2011), patents refer to technologies related to the electronic marketplace (orange cluster at the top of the tubes layout) and data management (three blue clusters in the third, fifth and sixth position of the tubes layout) (see Figure 5.10 in Appendix). In the second period (2012–2013), Amazon keeps as a priority to develop technologies improving its marketplace, but also starts developing a more distinct orientation towards AI and big data, within and beyond e-commerce inventions. The red cluster (at the top in the tubes diagram) reflects efforts to develop voice recognition and the use of voice commands for electronic devices, probably linked to Alexa's development, launched in 2014.

In this second period, the intermediate blue cluster points to attempts to establish technological competence in data management (storage and analysis). This cluster becomes more central in this period with terms such as "data store" and "data center" and it is tightly connected to the darker blue cluster on data analysis that includes the multi-term "virtual machine". All these patents reflect the ongoing development of technologies that, among others, support AWS, which in the third period

---

[12] https://academia.electronicsforu.com/microsoft-partners-with-alphabet-to-launch-online-course-on-quantum-computing.

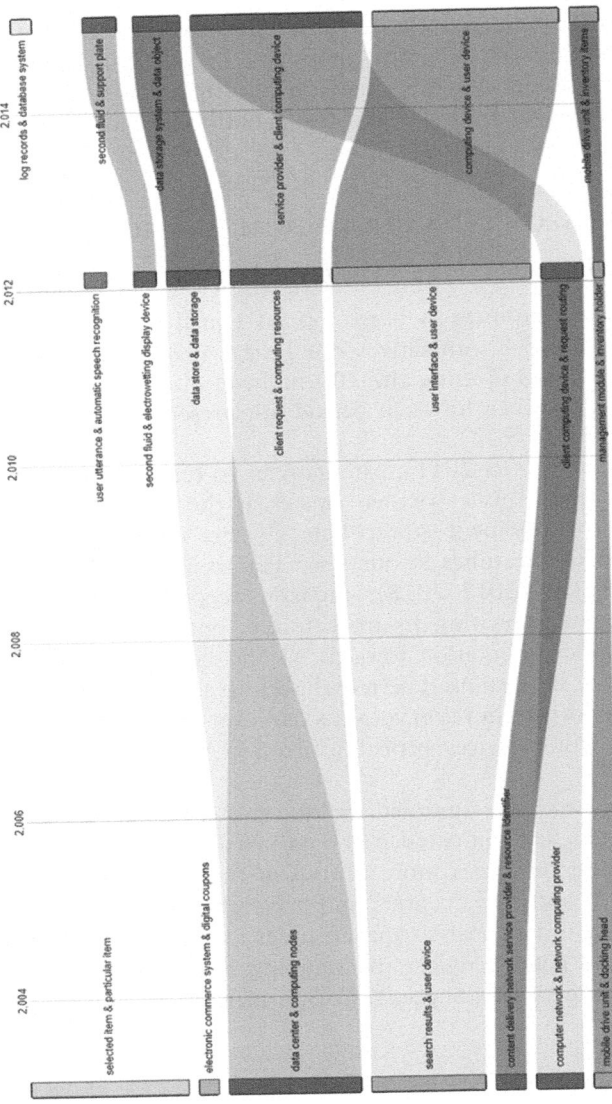

Fig. 5.4 Evolution of Amazon's patent portfolio content. Tubes' layout (*Source* Authors' analysis based on Derwent Innovation)

(2014–2017) becomes even more important—represented by the biggest cluster (intermediate blue).

Moreover, the small yellow cluster (at the top and to the right in the tubes diagram) that pops up in the third period also points to the relevance of AI and big data. It refers to interconnected patents on technologies for data storage and data retrieval. In this respect, the terms "storage node" and "server node" refer to Amazon's system Aurora. The company claims to offer a much faster mechanism to retrieve data stored in the cloud, with data better secured/encrypted.[13]

Overall, Microsoft's and Amazon's patent portfolios are focusing on the same technologies, even if they sometimes differ at the level of field applications. By analysing the content of Amazon's publications, we provide next more evidence of its technological convergence with Microsoft.

5.4.2 *Amazon's Recent Scientific Publications*

As in the case of Microsoft, we built two network maps, one with the 100 multi-terms with the highest frequency of co-occurrence in Amazon's scientific publications corpus between 2014 and 2017 (Fig. 5.5) and another one for the content of publications between 2018 and 2019 (Fig. 5.6).

The main cluster (in yellow in the middle of the network) in Fig. 5.5 deals with machine learning. It includes multi-terms related to research on specific techniques such as "natural language" and "sentiment analysis". The cluster refers both to more basic research on machine learning and to research applied to Amazon's different businesses. It also includes the term "recommendation systems" linked to the analysis of user review data. Machine learning algorithms are applied to topics beyond Amazon's current businesses, such as this cluster's paper "Burden of neurological diseases in the US revealed by web searches" (Baeza-Yates et al., 2017).

In line with Chapter 4 and as it was the case of Microsoft, Amazon's publications are dominated by research on deep neural networks. References to this topic appear in two clusters (one of the green ones and the

---

[13] Retrieved from https://aws.amazon.com/blogs/database/introducing-the-aurora-storage-engine/.

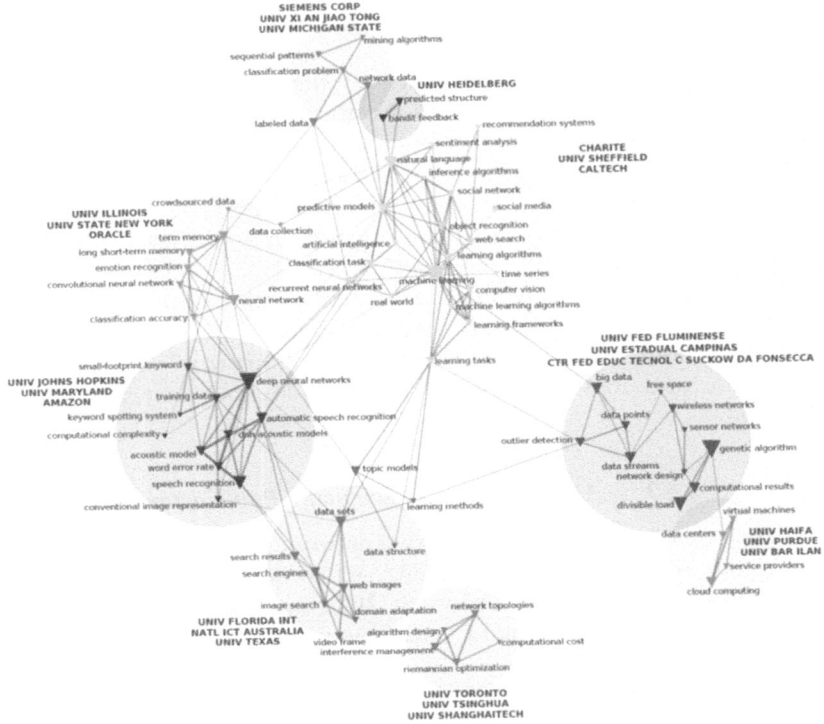

Fig. 5.5 Amazon's publications (2014–2017) lexical analysis (*Source* Authors' analysis based on Web of Science)

violet one), which are linked by the bridging multi-terms "deep neural network" and "neural networks". These are the two main bridging nodes of the map, highlighting their relevance in Amazon's publications, hence the importance of the AI TIS for Amazon's R&D strategy. The green cluster corresponds to research on more general techniques for emotion recognition and machine learning memory. The violet cluster is related to speech recognition. It is noteworthy that this cluster includes the term "conventional image representation", probably related to Echo Show,

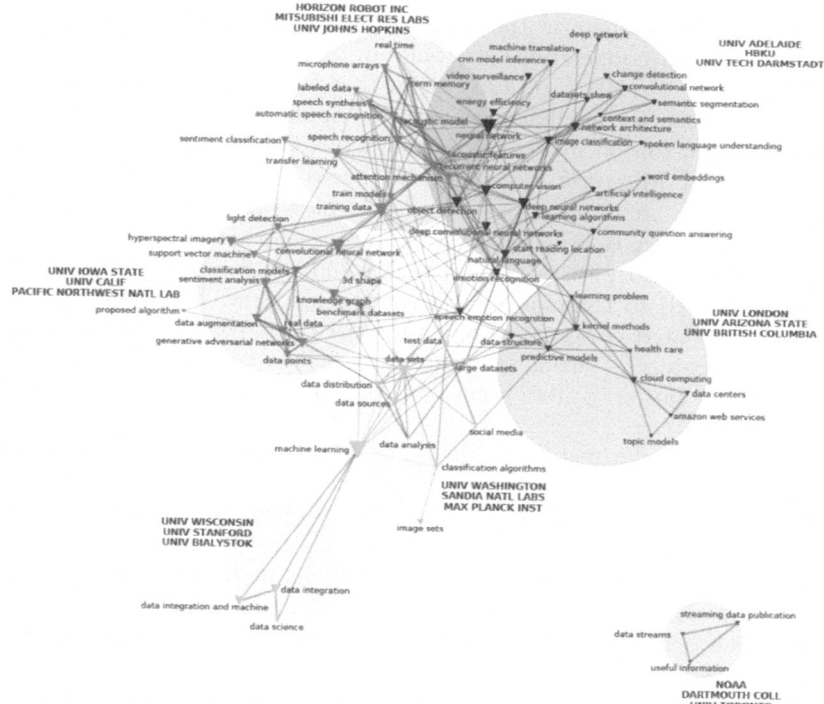

Fig. 5.6 Amazon's recent publications (2018–2019) lexical analysis (*Source* Authors' analysis based on Web of Science)

which is an Alexa speaker that includes a visual display.[14] Terms related to the data needed to train AI algorithms (which also appeared in the case of Microsoft) belong to this cluster.

The violet cluster includes a paper written by Amazon, Microsoft, Google and Rutgers University titled "The Future of Artificially Intelligent Assistants". This piece is quite telling of how these companies

[14] https://www.theverge.com/2017/6/26/15865292/amazon-echo-show-alexa-review.

simultaneously compete in the intelligent assistants' market and engage in technological cooperation, especially in general knowledge modules. Cusumano et al. (2019, p. 123) claim that each tech giant has found its market niche for intelligent assistants: "Apple focused on the quality of music; Amazon on media and e-commerce; Google on search-related inquiries; and Microsoft on enterprise needs". In every case, intelligent assistants are a source of big data—therefore a source of digital intelligence, hence of potential new businesses—for tech giants.

Figure 5.5 also plots a cluster in red and on the right side of the network related to techniques for analysing big data. This cluster has a smaller associated cluster (in orange) focused on AWS including terms like "data centers", "virtual machines" and "cloud computing". Another data-related cluster is the green one dealing with data mining and classification.

To sum up, the map is dominated by the AI TIS. Even the small cluster (in turquoise) detached from the network deals with these topics reflecting basic research on advanced mathematics for computing and network topology. Furthermore, in both content maps (Fig. 5.4 for patents' data and Fig. 5.5 for publications' data), multi-terms with "data" stand out (12 terms in the former and 10 in the latter). Both maps also depict a cluster integrated by web searches and cloud computing multi-terms like "virtual machines".

As in the case of Microsoft, we also built a network map with the 100 multi-terms with the highest frequency of co-occurrence in Amazon's publications in 2018 and 2019 (Fig. 5.6). This map shows that Amazon's publications remain focused on AI, particularly "machine learning" (yellow cluster) and especially on deep neural networks (biggest cluster in blue). Deep neural network techniques are applied to speech and emotions recognition as can be inferred from the light-blue and violet clusters. The light-blue cluster includes another bridge multi-term, "training data". Amazon is also developing machine learning models for real data (green cluster) and two other clusters reflect research on data. One is focused on data management (orange cluster) and the other one on machine learning through data science (yellow cluster). Finally, there is a small and detached cluster on raw data with the multi-terms "streaming data publications" and "data stream".[15]

15 https://aws.amazon.com/streaming-data/?nc1=h_ls.

A novelty of this period is that Amazon, as Microsoft (see Fig. 5.3), is researching on "video surveillance". Another novelty refers to healthcare digitalization. Between 2018 and 2019, Amazon's interest in health-care expands connected to its cloud business. This is reflected in the multi-term "health care" plotted directly linked to "cloud computing". Amazon claims that AWS offers "innovative cloud-based solutions to enhance patient care, lower costs, and improve efficiency".[16] Although here we found a difference with Microsoft, the latter's recent acquisition of Nuance, a cloud-based system for hospitals and doctors, points to another area of market and technological convergence in the near future (see Chapter 8).

All in all, and as we concluded for Microsoft, Amazon's publications are highly focused on data management and different AI techniques and applications to produce digital intelligence, pointing to the predominance of the AI TIS in Amazon's R&D priorities.

5.4.3 Concluding Remarks

Amazon and Microsoft have become vanguards of the digital revolution, with world-leading positions in AI development and cloud computing. In this chapter, we followed how they, with origin in very different economic and technological activities, have gone through science and technology transformations leading to technological convergence in the AI TIS. This is the technology that fuels all their businesses. We evidenced that the core of Amazon and Microsoft papers and patents evolved from topics focused on their original core businesses—thus differed—to deal with the same topics in recent years (in particular, from 2014 onwards). Their main converging dimensions are deep neural networks, data management and technologies for their cloud computing businesses.

Lexical analyses of patents and publications also provided clear insights on these two companies' focus on multiple AI functional applications using language processing (including sentiment analysis), speech recognition and object identification, including image recognition and video surveillance, all of which are ways to harvest more data, reinforcing

---

[16] https://aws.amazon.com/marketplace/solutions/healthcare.

their data-driven intellectual monopoly. Accordingly, our lexical analyses provided evidence of the centrality of harvesting, storing and processing data. As expected, given their leading position in the cloud computing market, cloud computing infrastructure is an area where the two companies' R&D overlaps. As we explained in Chapter 4, cloud computing is another example of a business that provides original streams of business data that enables tech giants to keep improving their AI models with third-party data.

These companies' technological convergence leads to converging in some markets—as was the case of cloud computing—but goes beyond their current businesses. By dominating the AI TIS, they could eventually expand to the most diverse—potentially every—industries (see Chapter 4). Given the relative weight of these digital companies, our analysis throws light on the formation of a new technological trajectory where, as we showed in Chapter 4, the critical trade-off is between data volume and machine learning. Yet, tech giants seem to be preparing to excel in both respects. This chapter showed the evolution of Amazon and Microsoft R&D strategy from their core businesses to focusing on machine learning and cloud computing, while Chapter 4 highlighted that these companies are also expanding their data volume by entering every AI application field.

In addition, our in-depth content analysis of the publications and patents of the two selected tech giants evidence how in line both type of R&D outcomes are. This is of particular importance given that, as we showed in Chapter 2, tech giants do publish with hundreds of other organizations but barely share the property of their patents. By evidencing that both patents and publications deal with the same technologies, we provide further evidence of tech giants' knowledge predation from their corporate innovation system.

We dedicate the two following chapters of this book to elaborate on the geopolitical dimension of this global innovation race. In Chapters 6 and 7 we argue that, to some extent, tech giants are taking over state responsibilities. While this poses challenges to states in general, it also reinforces the US and China's hegemony and the conflict between them taking the form of a (digital) technology war.

Appendix: Figures

See Figs. 5.7, 5.8, 5.9, 5.10, 5.11 and 5.12

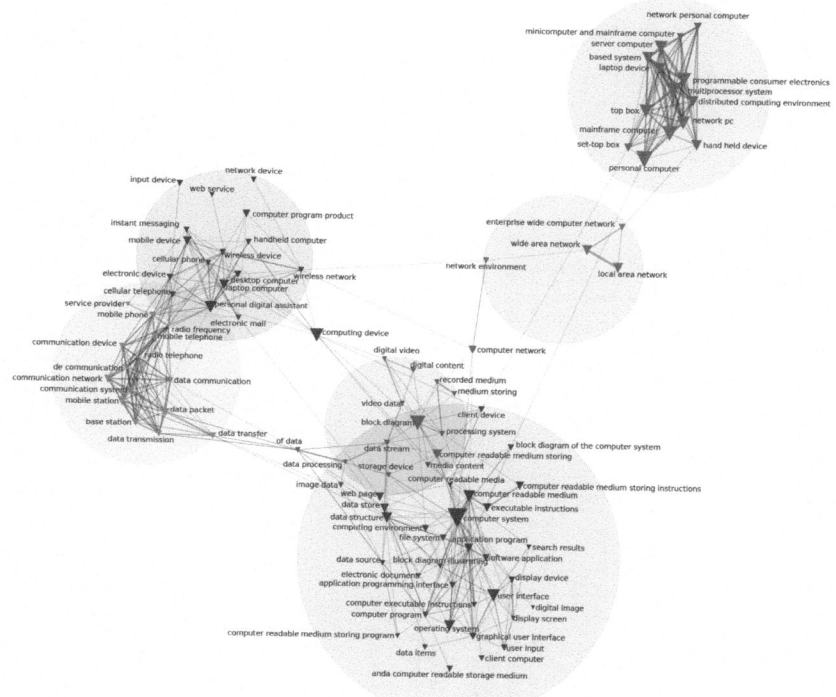

Fig. 5.7 Microsoft's patents. Lexical analysis (1986–2008) (*Source* Authors' analysis based on Derwent Innovation)

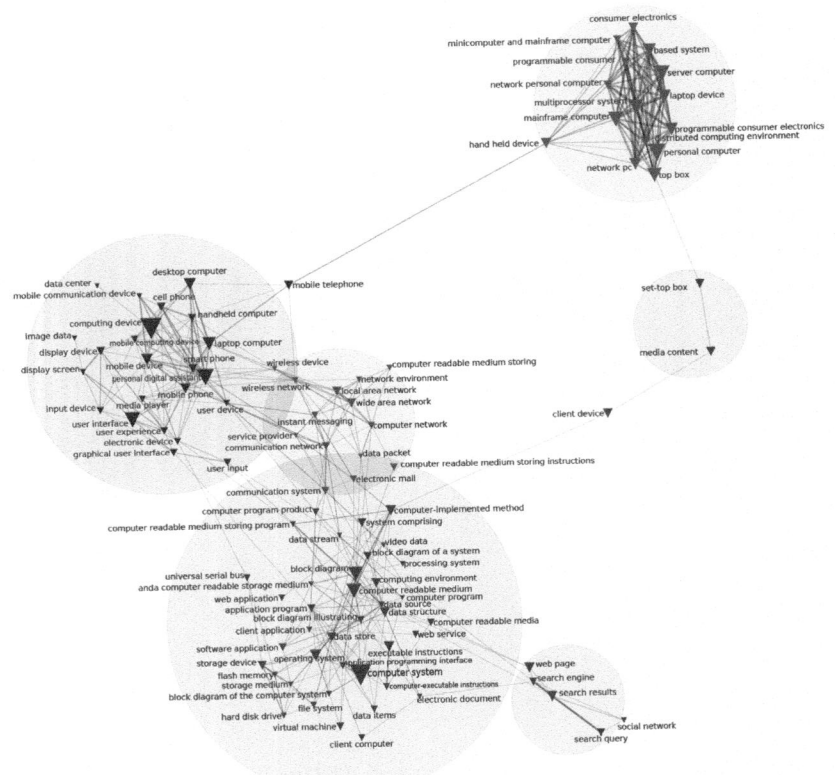

Fig. 5.8 Microsoft's patents. Lexical analysis (2009–2013) (*Source* Authors' analysis based on Derwent Innovation)

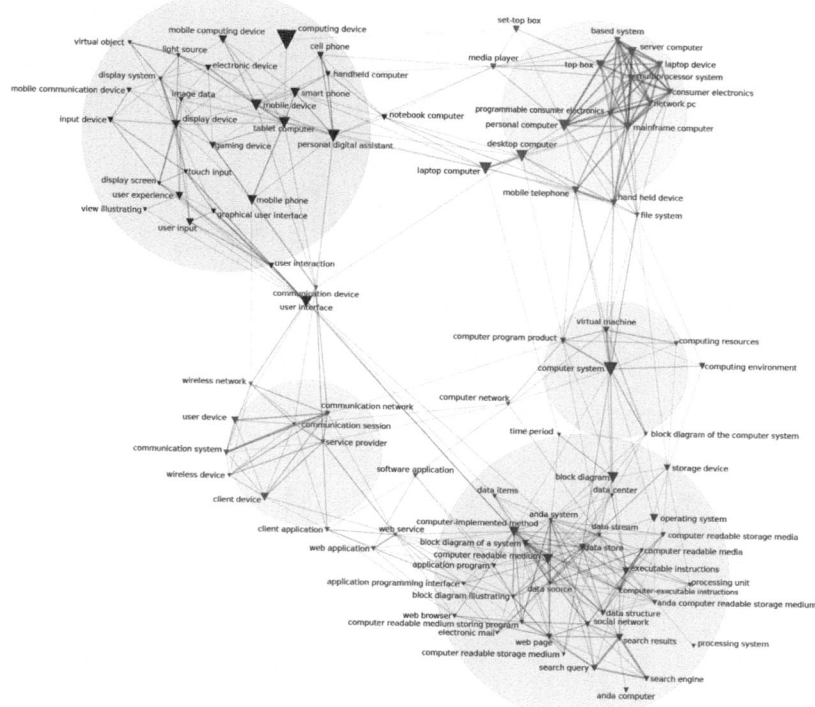

Fig. 5.9 Microsoft's patents. Lexical analysis (2014–2017) (*Source* Authors' analysis based on Derwent Innovation)

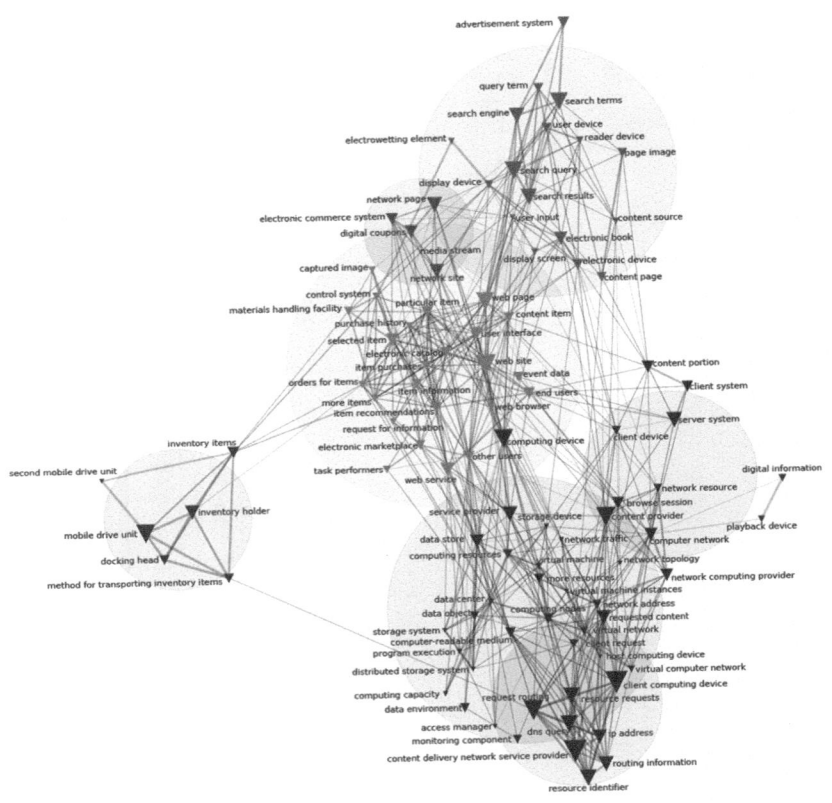

Fig. 5.10 Amazon's patents. Lexical analysis (1996–2011) (*Source* Authors' analysis based on Derwent Innovation)

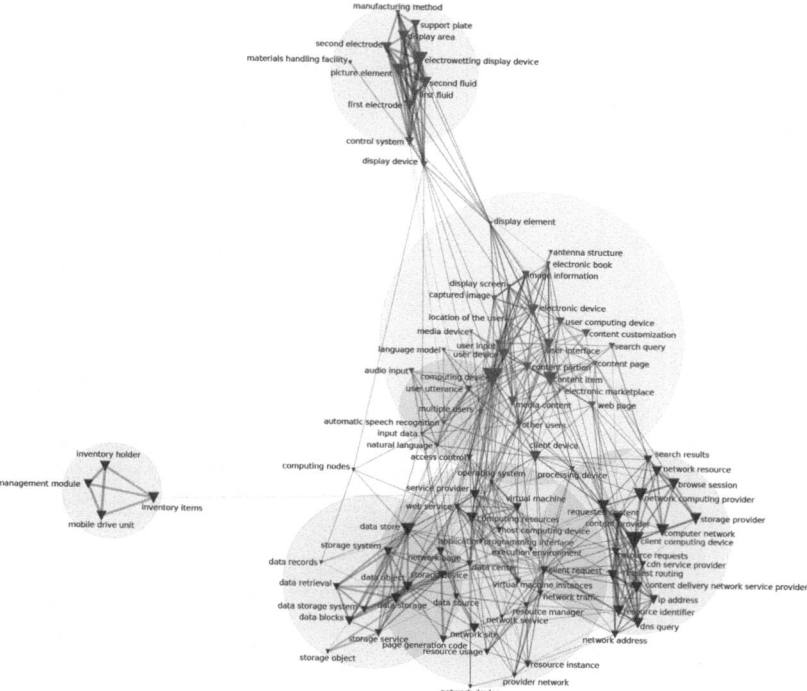

Fig. 5.11 Amazon's patents. Lexical analysis (2012–2013) (*Source* Authors' analysis based on Derwent Innovation)

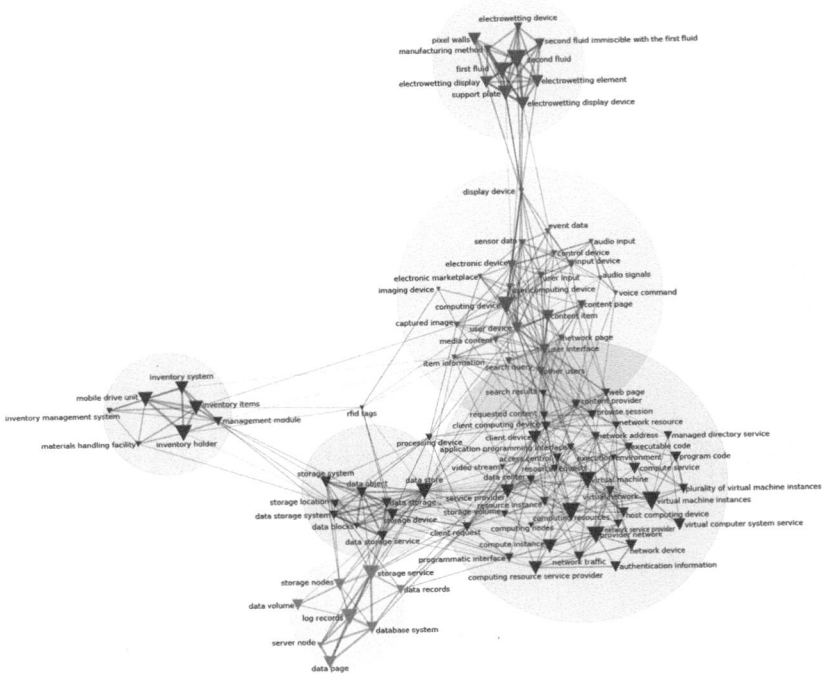

Fig. 5.12 Amazon's patents. Lexical analysis (2014–2017) (*Source* Authors' analysis based on Derwent Innovation)

REFERENCES

Arthur, C. (2012). *Digital wars: Apple, Google, Microsoft and the battle for the Internet.* Kogan Page Publishers.

Baeza-Yates, R., Sangal, P. M., & Villoslada, P. (2017). Burden of neurological diseases in the US revealed by web searches. *Plos One, 12*(5), e0178019.

Choi, S., Yoon, J., Kim, K., Lee, J. Y., & Kim, C.-H. (2011). SAO network analysis of patents for technology trends identification: A case study of polymer electrolyte membrane technology in proton exchange membrane fuel cells. *Scientometrics, 88*(3), 863.

Cusumano, M. A., Gawer, A., & Yoffie, D. B. (2019). *The business of platforms: Strategy in the age of digital competition, innovation, and power.* HarperCollins Publishers.

Dosi, G. (1981). Technical change and survival: Europe's semiconductor industry. *LEM Book Series*.

Dosi, G. (1982). Technological paradigms and technological trajectories: A suggested interpretation of the determinants and directions of technical change. *Research Policy, 11*(3), 147–162.

Fortunato, S., & Hric, D. (2016). Community detection in networks: A user guide. *Physics Reports, 659*, 1–44. https://doi.org/10.1016/j.physrep.2016.09.002

Ibarra, H., & Rattan, A. (2018). Microsoft: Instilling a growth mindset. *London Business School Review, 29*(3), 50–53.

Khan, L. (2017). Amazon's Antitrust Paradox. *The Yale Law Journal, 126*(3), 710–805.

Madani, F., & Weber, C. (2016). The evolution of patent mining: Applying bibliometrics analysis and keyword network analysis. *World Patent Information, 46*, 32–48.

Pisano, G. P. (2015). You need an innovation strategy. *Harvard Business Review, 93*(6), 44–54.

Rikap, C. (2020). Amazon: A story of accumulation through intellectual rentiership and predation. *Competition & Change*. https://doi.org/10.1177/1024529420932418

Srnicek, N. (2017). *Platform capitalism*. Wiley.

Testoni, F. E., García Carrillo, M., Gagnon, M.-A., Rikap, C., & Blaustein, M. (2021). Whose shoulders is health research standing on? Determining the key actors and contents of the prevailing biomedical research agenda. *PloS One, 16*(4), e0249661.

UNCTAD. (2019). *Digital Economy Report 2019: Value creation and capture–Implications for developing countries*. United Nations.

Van Looy, B., & Magerman, T. (2019). Using text mining algorithms for patent documents and publications. In *Springer handbook of science and technology indicators* (pp. 929–956). Springer.

World Intellectual Property Organization. (2019). *WIPO technology trends 2019: Artificial intelligence*. WIPO.

Yoon, J., Choi, S., & Kim, K. (2011). Invention property-function network analysis of patents: A case of silicon-based thin film solar cells. *Scientometrics, 86*(3), 687–703.

CHAPTER 6

China's Catching-Up Process and Its Emergence as a Potential Lead Country in Artificial Intelligence

Abstract This chapter explores China's catching-up in artificial intelligence (AI) with high-income countries. It is inspired by Christopher Freeman's work on, how radical technical change opens up for shifts in world leadership. Focusing on artificial intelligence as a core technology of the second phase of the Information and Communication Technology revolution, we analyse the strengths and weaknesses of China's innovation system focusing on its catching-up in AI, seen as a co-evolution of the country's national innovation system and the corporate innovation systems of Alibaba and Tencent.

Keywords Catching-up · Corporate Innovation System · National Innovation System · Christopher Freeman · China · Big tech

6.1 INTRODUCTION

An apparent feature of the twenty-first century is China's turn from the factory of the world into a digital economy leader just behind the US, while Europe and Japan are laggards (UNCTAD, 2019). Why do some countries catch up while others fall behind? What is the basis

C. Rikap and B.-Å Lundvall, *The Digital Innovation Race*, https://doi.org/10.1007/978-3-030-89443-6_6

of world hegemony and under what circumstances and how can late-comer countries forge ahead and, eventually, become world leaders? Do these long-standing questions have new answers in contemporary digital capitalism?

According to Christopher Freeman, answering these types of questions involves considering the uneven (but not completely irregular) development of technology and the national innovation system (NIS). Freeman (2007) linked shifts in global leadership to technological revolutions. He explained that, in the eighteenth century, Britain's NIS had developed characteristics (and systemic coherence) contributing to explain why it became the homestead for the industrial revolution based upon steam power and textiles. As new technological systems dominated by electricity and chemistry emerged, Germany and the US forged ahead and left England behind. Freeman's (1987) analysis of Japan's emergence as a potential technological leader in an era of information technology illustrated this general hypothesis.

Picking up from Freeman's research, Pérez and Soete (1988) argued that major technological changes create windows of opportunity where leapfrogging is possible because the incumbents' investments are locked into older technologies leaving the erstwhile laggards free to adopt the new technologies. Keun Lee (2013) has developed the analysis of leapfrogging and technological catching-up highlighting the importance of entering into "short-cycle technologies"—operationalized as technologies with a short history of patenting.

Does AI and related technologies represent such an opportunity for catching-up? While AI centrality is not a matter of discussion, there is still a debate in terms of its revolutionary scope. World Economic Forum (Schwab, 2016) argues that AI, together with biotechnology, is at the core of a fourth industrial revolution while others, such as Nuvolari (2019) and Brynjolfsson and McAfee (2014), see the new developments as taking place within the third industrial revolution. Jongho Lee and Keun Lee (2021) attempt to solve this debate using USPTO patent data. Based on an analysis of the generality, radicality and length of technology cycles, they conclude that there is no clear indication of a radical break in technology.[1]

---

[1] Applying the length of technology cycles using patent history to AI may underestimate the fact that AI, while having a long historical record of patenting, has gone through a radical transformation since 2012, when specific processors for deep learning algorithms

Here we refer to the ICT revolution as constituted by two phases with AI at the core of the second phase and we assume that AI will have the same transformative potential now that microelectronics had in the first phase of the ICT revolution. We therefore propose that it offers a window of opportunity for China similar to the one that Japan exploited in the 1970s and 1980s. On this basis, we argue that China's success in catching-up will depend, among others, on the strength of its NIS and on how it can support building AI capabilities—spanning from basic research to end-users.

Nevertheless, in order to understand catching up in AI, we need to go beyond analysing the NIS. The emergence of (data-driven) intellectual monopolies (see Chapter 2) represents a major change in current capitalism. While they draw upon their respective domestic NIS, they simultaneously organize and lead innovation at the global level through Corporate Innovation Systems (CIS) (see Chapter 3). We thus propose to understand China's AI catching-up process as reflecting co-evolution[2] between NIS and a few tech giants' CIS. Therefore, in this chapter we integrate political economy insights on the interplay between corporate and political powers with innovation studies' contributions.

The chapter is organized as follows. In Sect. 6.2, we highlight some features of China's NIS focusing on its catching up in AI. Section 6.3 elaborates on the coevolution between the Chinese state and China's tech giants, in particular considering the experience of Alibaba and Tencent. Sect. 6.4 concludes.

---

were introduced. Recent "rejuvenation" and acceleration of AI reflect a combination of rapid progress in complementary technologies and radical social innovations. The explosive growth of data harvesting through major platforms contributes directly to the acceleration of the innovation process.

[2] We borrow the term *co-evolution* from biological sciences where it refers to the "the process of reciprocal evolutionary change that occurs between pairs of species or among groups of species as they interact with one another". This means that their respective activity affects each other's evolution (https://www.britannica.com/science/coevolution). In our case, different groups of species organized in communities (the NIS and the CIS) co-evolve.

6.2   CHINA'S NATIONAL INNOVATION SYSTEM AND ITS CATCHING UP IN AI

China's innovation system is large, heterogenous and utterly complex. Regions are very different in terms of innovation capabilities and innovation modes. While the central state and the top leaders of the communist party set the general direction for innovation policy, the provinces and lower-level local authorities have a certain autonomy in executing policy, including STI-policy (Gu & Lundvall, 2006a, 2006b). Moreover, huge state-owned enterprises (SOEs) co-exist with private firms and, as illustrated by Baidu, Alibaba, Tencent and Huawei, entrepreneurship has resulted in the creation of giant corporations operating globally (Liu et al., 2021).

One way to reduce complexity is by presenting aggregate quantitative variables on innovation inputs and outputs. As we show in Sect. 6.2.2 below, indicators show that in both these dimensions China's innovation system has gone through a very rapid expansion over the last forty years.

6.2.1   Brief Remarks on the Evolution of China's NIS

In this sub-section, we refer to three aspects of the evolution of China's NIS that we deem to have major relevance for the development of AI in China: (1) the role of state-owned enterprises (SOE), (2) coupling between knowledge institutions and industry and (3) the openness of the system.

Concerning the first aspect, the SOE's share of the national economy remains substantial. According to Zhang (2019), they constitute around 17% of industrial employment, while their role in promoting or hampering innovation remains controversial. Some scholars conclude that they contribute more to innovation than other types of enterprises (Qi & Kotz, 2020). However, Liu et al. (2017) demonstrate that they absorb most of the government's support for business R&D and point out that China's most successful big enterprises have grown into global (tech) giants with little or no government support. Nonetheless, the State Grid Corporation of China (SGCC), a SOE, is China's top organization in AI patenting (World Intellectual Property Organization, 2019) and it heavily relies on knowledge produced by Chinese universities for its R&D

(Rikap, 2021, Chapter 9). Furthermore, the recently introduced regulations of fintech businesses directly affected Alibaba and Tencent and favoured state-owned banks (see Sect. 6.3).

Regarding the second aspect, as early as 1985, the then prime minister Zhao Ziyang diagnosed the weak interaction between knowledge institutions and industry as a major problem of the Soviet model for organizing research, with sector-specific research institutes weakly connected to enterprises. In the 1980s, some of China's ensuing attempts to create markets for knowledge and to force research organizations to sell knowledge directly to firms failed. Instead, research organizations engaged in establishing their own enterprises (Eun et al., 2006; Gu & Lundvall, 2006a). According to Liu et al. (2017), this crucial dimension of the NIS remains problematic. For instance, while elite universities actively apply for patents, only a small share is licensed by enterprises or otherwise commercialized.

Another indicator of weak couplings is the small and stagnating share of enterprise R&D allocated to university research, which should be considered in comparison with similar figures for other countries (see Fig. 6.1). These results could be seen as reflecting a similar mechanism as the one discussed in Chapter 3, i.e. higher education institutions' difficulty to profit from their innovation, suffering knowledge predation from tech giants. Overall, and in line with higher education systems in the rest of the world, polarization between leading research universities with strong ties with the industry (as the University of Science and Technology of China or Tsinghua University, see Chapter 3) and a big number of institutions focusing on teaching and positioned far from the world rankings points to the limits of analysing university–industry links with average figures (Rikap & Harari-Kermadec, 2019) (Fig. 6.1).

The third aspect, the opening of the economy, is especially interesting seen from the NIS perspective. Freeman's (1982, 1987) different contributions to the analysis of NIS performance have focused on this dimension. He argued that one major explanation of Japan's success was that it invested heavily in domestic knowledge and learned from abroad through reverse engineering, without opening up for foreign direct investments (FDI).

China has followed a different strategy opening up for FDI in the 1980s, aiming to learn directly from foreign firms. The government used

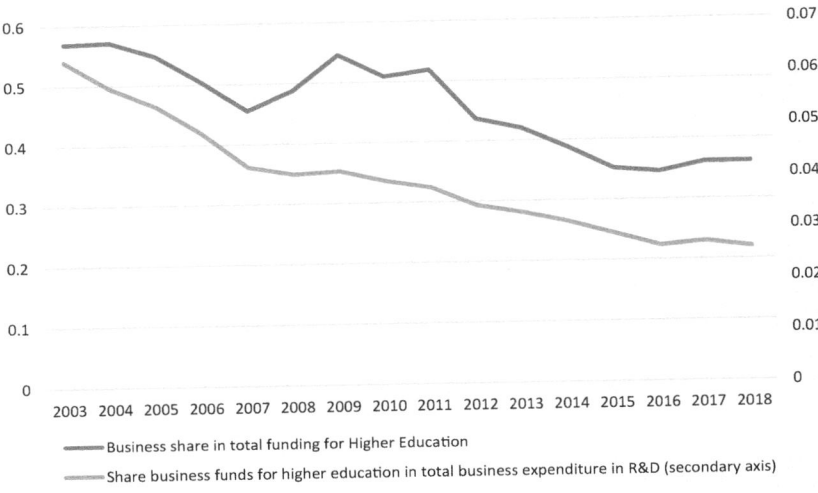

0.6   0.07
0.5   0.06
                                                                       0.05
0.4   0.04
0.3   0.03
0.2   0.02
0.1   0.01
0   0

2003 2004 2005 2006 2007 2008 2009 2010 2011 2012 2013 2014 2015 2016 2017 2018

━━━ Business share in total funding for Higher Education

━━━ Share business funds for higher education in total business expenditure in R&D (secondary axis)

Fig. 6.1   Business expenditure in R&D allocated to Higher Education (*Source* Author's analysis based on OECD data)

foreign multinational firms' desire to access China's huge domestic market and low real wages to negotiate access to their knowledge through joint ventures. The success of this strategy was mixed. For a long period, high growth rates were based on low wages and high rates of capital accumulation rather than on technical progress.

In an OECD working paper, Schaaper (2009) praises the role of multinational enterprises in technological upgrading in China stating that "the catch-up in high-technology outputs and exports is largely attributable to inward FDI". At the same time, the data presented in the report points to the limits of the openness strategy. By 2005, most of China's high-technology exports came from foreign-owned enterprises importing components and using labour-intensive processes to assemble final products. The report also showed that Chinese accumulated triadic patent families remained extremely low—433 as compared to around 15,000 in the US, Europe and Japan. The 2006 policy aiming at indigenous innovation may be seen as a reaction to the weaknesses reflected in such indicators (Gu & Lundvall, 2006a).

According to Feng (2019), the lack of success in building technological capabilities reflected the combination of unwillingness of foreign partners to share technology and the lack of incentives for SOE management. Since there was no competition from indigenous producers, multinationals harvested huge profits from the Chinese market, often through the sales of products based upon outdated technologies.

The strategic shift from imitation to "independent" (alternatively translated as indigenous) innovation was implemented with the 2006–2020 Plan for the Development of Science and Technology in the Medium and Long Term, where the term "indigenous innovation" became the keyword. A crucial ambition of this plan was to overcome the dependence of foreign firms in advanced technologies. This new direction towards indigenous innovation was confirmed in the Made in China 2025 (MiC) plan and again in China's 13th Five-Year Science and Technology Plan and in its most recent five-year plan (2021–2025).

This strategy implied that the openness of the economy should be managed differently. Chinese enterprises had already engaged in a different form of openness through outward FDI since 1999, encouraged by the Chinese government to "go global". In 1986, China accounted for 0.1% of global outward FDI stock. This share rose to 0.48% by 2001 and reached 4.9% in 2016 (Buckley et al., 2018). Originally focused on increasing access to resource-based commodities, outward FDI has recently changed towards access to technological learning. According to Fu et al. (2018), reverse learning—where Chinese enterprises learn from customers, collaborators and affiliates in developed economies—was of major importance, in particular, for Huawei and ZTE.

Altogether, the three dimensions reviewed here are important for understanding China's STI-performance as presented in the next subsection and they also impact on the development and use of AI in China addressed in Sect. 6.2.3.

6.2.2 On the STI Performance—Quality vs Quantity

Since the beginning of the 1990s, China has seen extremely high growth rates in investment in different forms of knowledge as well as in certain knowledge output indicators.

In two decades, graduates from Chinese universities grew tenfold, reaching 7 million students in 2017, more than doubling US figures (World Bank & DRC State Council, The People's Republic of China, 2019). R&D expenditure as a share of GDP also exhibits impressive

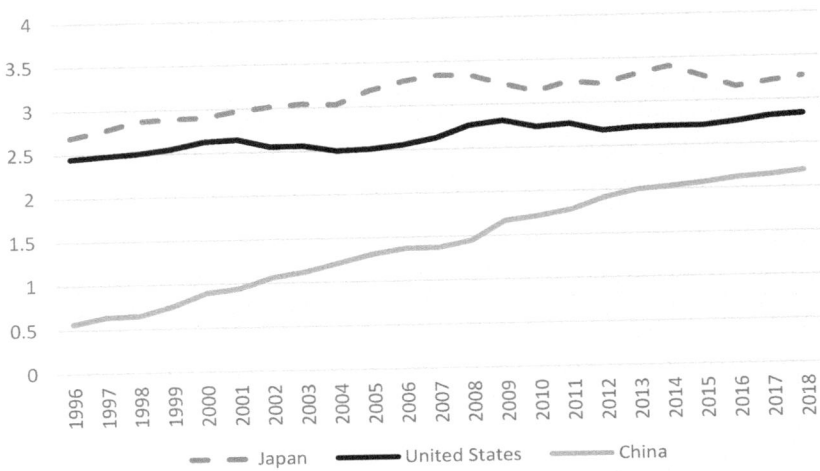

Fig. 6.2 R&D expenditures as a share of GDP. Selected countries (*Source* World Bank)

results (Fig. 6.2) and it has now reached the level of the European Union (2.1%).[3]

As shown in Table 6.1 and Fig. 6.3, China's output growth in terms of scientific papers and patents has accelerated. China is now the world leader both in terms of the number of scientific publications and patents.

The rapid growth in investments in knowledge, as well as the growth in outputs, needs to be critically assessed in the light of incentive structures. To access public funding from technology programs, enterprises must document a certain volume of R&D and patenting activities (Tseng, 2009; Zhou & Stembridge, 2008). At universities, both research grants and salaries depend on the number of publications and patent applications (Liu et al., 2017). As agents respond to these incentives, it results in inflated numbers and weak quality of both publications and patents.

The focus on numbers and quantitative success indicators, which neglects harder to observe but crucial dimensions of the innovation system, results in a weakness of China's innovation policy. Weak quality

[3] See https://ec.europa.eu/eurostat/statistics-explained/index.php/R_%26_D_expenditure.

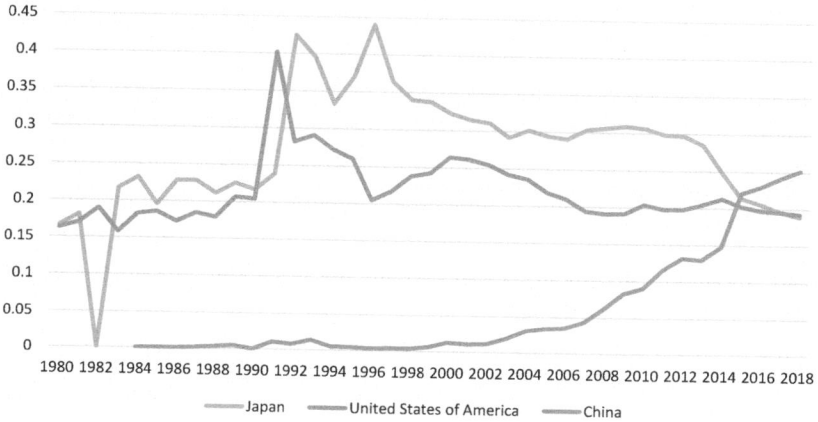

Fig. 6.3 Share of total granted patents. Selected countries (*Source* Authors' analysis based on WIPO)

Table 6.1 Publication output.[4] Selected countries

	2008	2018	Average annual growth rate 2008–18 (%)	2018 world total (%)
China	249,049	528,263	7.81	20.67
United States	393,979	422,808	0.71	16.54
Japan	108,241	98,793	−0.91	3.87

Source US National Science Foundation

controls through collegial evaluation systems contribute to this trend (Fu et al., 2016; Liu et al., 2017). Such weakness is reflected in the fact that Chinese organizations are less successful than foreign ones in granted to applied patents' ratio and still mostly apply for patents only inside China (China Clarivate Analytics, 2018). Only 6% of Chinese patents are taken abroad, while this share is 30% for US applicants' patents (Benoliel, 2017).

---

[4] Data on publication output includes publications from peer-reviewed journals and conference proceedings in science and engineering and indexed in Scopus. One count was assigned to each country or institutional sector involved in co-authoring the article, irrespective of their proportionate involvement in authorship.

China's catching-up in production methods remains incomplete. The US, the EU, Japan and South Korea are well ahead in several fields such as automobiles and semiconductors (Lee et al., 2017). Nonetheless, Chinese companies are moving beyond imitation and towards an innovation-driven stage in some sectors and technologies (Gu et al., 2016). Firms operating in China are increasingly involved in complex production processes. National Science Foundation data shows that China's share of the world's R&D intensive industries grew from 6% in 2003 to 21% by 2020, second after the US, whose share fell from 38 to 32% (National Science Foundation, 2020).

Chinese firms are now frontrunners in high-speed trains, wind turbines, supercomputers, space launch vehicles and satellites and liquid crystal displays (LCDs) (Godinho & Ferreira, 2012; Guo et al., 2019; Hawes & Chew, 2011; Humphrey et al., 2018; Li et al., 2019; Li & Cheong, 2017; Wen, 2017; Wu & Gereffi, 2018).

Furthermore, different signs are pointing to China achieving fore-runner status in 5G, AI and quantum computing[5] (Shi-Kupfer & Ohlberg, 2019). Scherngell et al. (2020) showed that China is occupying a central place in the network of international co-invented patents in telecommunications—primarily explained by Huawei's 5G leadership. Since the frequency of international collaborations is a sign of higher R&D quality, the authors put into question the doubts about the quality of China's ICT patents. In the next section, we elaborate on China's catching up in AI.

6.2.3   China's Catching-Up in Artificial Intelligence

In this subsection, we use World Bank, WIPO and OECD data and reports on science, technology and innovation to compare China's NIS AI capabilities with those of the US and other core countries. In addition, we retrieved from the Web of Science two corpora. One includes all the publications with the term "machine learning" and the other all the publications with the terms "neural network" or "deep learning" both for 2019. We used these data to proxy the US and China's respective positions in these technologies.

---

5 See https://www.nature.com/articles/d41586-020-03434-7.

We also considered Allen's (2019) and Ernst's (2020) fieldworks in China. The former included interviews and participative observation in meetings with high-ranking Chinese officials, while the latter mainly interviewed Chinese enterprises. Moreover, our findings were validated with informal talks with Chinese scholars and AI developers working for Chinese tech companies.

We will argue that the role of the Chinese state and, in particular, the setting and strengthening of China's NIS have been crucial for China's ongoing AI catching-up. We also acknowledge that China's AI research mostly relies on its universities and public research organizations. Chinese corporations, in particular tech giants, have profited from these institutions engaging less in AI research as compared to the US tech giants. Overall, our analysis outlines that AI catching up is an ongoing and uneven process where China is already among leaders yet behind the US.

6.2.3.1   Strengths of China's AI Catching-Up

A recent report of the Center for Data Innovation found that China is the world leader in AI adoption and data (Castro & McLaughlin, 2021) and different reports have shown that China arrives first in AI publication counts (Castro & McLaughlin, 2021; China Institute for Science and Technology Policy at Tsinghua University, 2018; Stanford, 2021; Zhang et al., 2021). The Chinese Academy of Sciences is the world's first organization in AI publishing with over 26,000 publications until 2017 included and the second in terms of highly cited papers.[6]

Moreover, from a total of 38,224 publications with the term "machine learning", we found 11,113 (29%) written by at least one US-based author and 8502 (22%) by at least one China-based author. Similar concentration rates with the US and China as absolute leaders, but with China at the top, can be observed for publications including the terms "deep learning" or "neural networks". By 2019, from a total of 66,202 publications including either term, China had co-authored 23,186 (35%) and the US 14,685 (22%).

The Chinese state has played a central role in China's AI successes. It has funded AI research at universities and public research institutes and invested heavily in telecommunication infrastructure since the turn of the millennium (Hong, 2015). Moreover, a stepping stone in China's

---

[6] Defined as papers that perform in the top 1% based on the number of citations received within a field for a selected year.

AI catching-up was further developing China's Great Firewall, originally introduced in 1997. It consists of an ensemble of social media regulations, IP blacklists, keyword filters, data gateways and human censors (Tsui, 2007). The Firewall limits access for foreign companies to internet-based business (including cloud services), which favoured local companies in particular Baidu, Alibaba and Tencent (Azmeh et al., 2020; Mueller, 2011; Wu & Gereffi, 2018).

The Chinese state AI-related policies also include the Digital Silk Road, a new internet traffic route from China to Europe exclusively managed and supervised by the Chinese and announced in 2015 as part of the Belt and Road Initiative (BRI). It consists of cross-border cables connecting the BRI participating countries and Europe. By early 2021, Huawei and the Pakistani government were finishing the installation of the fibre optic cable in Pakistan, which by March 2021 was supposed to be connected to a submarine cable in the Arabian Sea creating the "Pakistan East Africa Connecting Europe" (PEACE) cable in the Indian Ocean and the Mediterranean, therefore finalizing China's Digital Silk Road.[7]

These policies, together with the size of China's population (1.3 billion) and the fast adoption of mobile phone usage (growing from 300 to 900 million users between 2010 and 2020[8]), internet and 4G resulted in a huge amount of data produced and stored inside China. China's datasphere is growing 3% faster than the global datasphere[9] and by 2018 China's total data volume was 23.4% (7.6 zettabytes) of the world's data volume (International Datacenter Corporation, 2019).

All the latter enabled Chinese tech giants to harvest enormous volumes of data—a prerequisite for establishing themselves and China as world leaders in AI applications. China follows the US, with 22% of the value of the world's 70 largest platforms (UNCTAD, 2019). China also comes second after the US in terms of share in the top 100 unicorns by value (PWC, 2020) and AI companies. As we mentioned in Chapter 4, by June 2018, there were 4925 AI enterprises worldwide, 42% from the US and 20% from China (China Institute for Science and Technology Policy at Tsinghua University, 2018).

---

[7] https://asia.nikkei.com/Spotlight/Belt-and-Road/China-builds-Digital-Silk-Road-in-Pakistan-to-Africa-and-Europe.

[8] https://www.statista.com/statistics/273973/number-of-mobile-internet-users-in-china/.

[9] The global datasphere is all the data created and replicated in one year.

Two decades after the introduction of the Great Firewall, Alibaba and Tencent are digital forerunners in e-payment and other fintech solutions. They are also among the world-leading companies in multiple digital services and platforms. As we show in Sect. 6.3, Chinese tech giants also rely on the AI R&D capabilities of Chinese universities and public research organizations, where AI research is mostly concentrated and where the most active AI patent applicants come from.

These results are indicative of a broader technological catching-up of China's NIS in AI, as evidenced by the overall evolution of its AI patent portfolio (see Fig. 6.4).

Behind these impressive results in patenting and scientific publications, public knowledge institutions seem to have a more active role

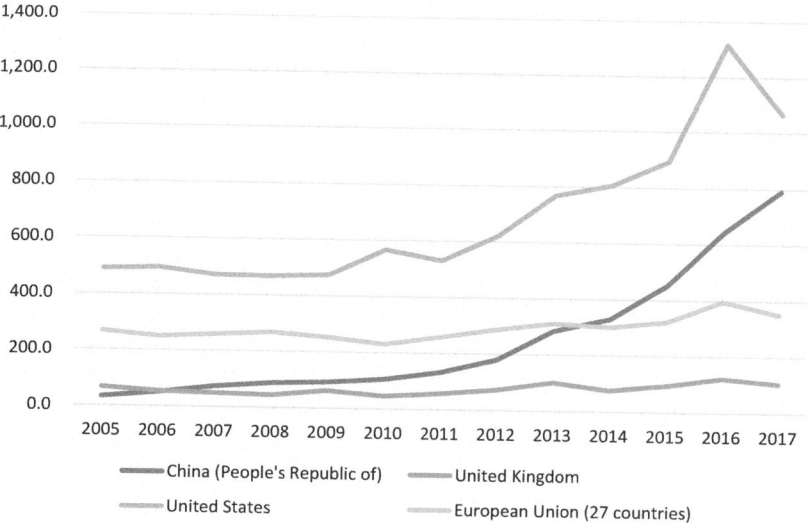

Fig. 6.4 AI patent counts by inventor(s)'s country(ies) of residence[10] (*Source* OECD data based on IP5 Patent families)

[10] Inventions protected in at least two jurisdictions, at least one of which needs being one of the top Five IP Offices.

than in the US. Most of China's top patent owners are public organizations (China Institute for Science and Technology Policy at Tsinghua University, 2018).

6.2.3.2   Weaknesses of China's AI

In terms of weaknesses, the Chinese leadership is particularly concerned with AI top talent, technical standards, software platforms and semiconductors (Allen, 2019). These weaknesses were emphasized in China's white paper on "Artificial Intelligence Standardization", written between public actors and experts from Baidu, Alibaba, Tencent, Huawei and Bytedance.[11]

To tackle these weaknesses, China's Next Generation AIDP gives priority—among others—to setting AI international standards. To accelerate the creation of Chinese global AI leaders, the plan advocated for helping AI firms to strengthen their patent structures and take the lead or, at least, participate in the formulation of international standards. It also claims that AI standards' setting should contribute to strengthening military–civilian integration in AI and accelerating AI adoption in countries along China's Belt and Road Initiative. As a follow-up of the AIDP plan, the mentioned white paper on "Artificial Intelligence Standardization" reiterated that special efforts to contribute to international standards were a prerequisite for accomplishing the AIDP goals.[12]

Concerning AI talent, by 2017, Europe more than doubled China in the number of AI researchers (43,064 versus 18,232, respectively) and the US had 56% more AI researchers than China (Castro et al., 2019). Nevertheless, aggregated data from 2015 to 2020 shows that China's AI skill penetration[13] rate is 1.40 times the global average. In this indicator, China ranks third after India (2.83) and the US (1.99) (Zhang et al., 2021). It may thus be the case that China's AI-talent demand is growing faster than internal AI-talent supply because of the accelerated but uneven AI catching-up of the country.

---

[11] https://www.newamerica.org/cybersecurity-initiative/digichina/blog/chinese-interests-take-big-seat-ai-governance-table/.

[12] https://www.newamerica.org/cybersecurity-initiative/digichina/blog/translation-excerpts-chinas-white-paper-artificial-intelligence-standardization/.

[13] Shows the average share of AI skills among the top 50 skills in each occupation, using LinkedIn data on skills by member considering positions held and the locations of the positions (Zhang et al., 2021).

Within semiconductors, leading-edge AI chips are critical for processing big data with machine learning (in particular using deep learning). China lags behind leading semiconductor firms whose chips assure computing power at lower costs (Ernst, 2016). Chinese firms occupy a marginal position in the overall chips value chains, including AI chips. Lagging in AI processing power is a fundamental weakness because of its centrality for AI development. In fact, a recent investigation found that "the compute divide", defined as uneven access to computing power, could be a major reason underlying the concentration of AI research by large technology firms (Ahmed & Wahed, 2020).

Chips are of paramount importance both for the US and China, but it may be argued that the former has an advantage. Although Taiwan's TSMC and South Korea's Samsung are the world leaders in production capacity, with a clear advantage of the former, the US strength comes from the impossibility to produce advanced semiconductors without access to leading US players' design technology (see Chapter 7).[14]

Ernst's (2020) in-depth fieldwork found that this is forcing China to accelerate AI catching-up in core foundational technologies with many AI chip initiatives on their way. As part of its AIDP plan, the Chinese government subsidized two of China's lead AI chips firms—SMIC and Tsinghua Unigroup—with over 30% of their respective annual consolidated revenue. However, according to Ernst (2020), SMIC—which seems to be the most advanced—would require at least a decade to close its gap with TSMC.

In terms of other weaknesses, insufficient early-stage R&D has been highlighted in the Chinese government AI megaprojects' guide, referring both to a need to foster fundamental theories of AI and key technologies (Colvin et al., 2020). Differences with the US point to the lower engagement of the private sector. While academia ranks first in AI publishing in both the US and China, corporations are the second most important AI paper author in the former (19.2% of total publications), whereas in China the government occupies the second position (Zhang et al., 2021). Only SGCC belongs to the world's top 20 enterprises in AI paper output occupying the 14th position. This ranking is led by Microsoft and IBM (China Institute for Science and Technology Policy at Tsinghua University, 2018).

---

[14] https://asia.nikkei.com/Spotlight/The-Big-Story/US-China-tech-war-Beijing-s-secret-chipmaking-champions.

Moreover, most of China's top patent owners are non-firm organizations. The only exception is SGCC, which is China's first AI patent holder with 4246 AI patents between 2013 and 2017. In the same period, the global leader IBM was assigned 7276 AI patents (China Institute for Science and Technology Policy at Tsinghua University, 2018).

Castro et al. (2019) found that China arrives after the US and the European Union when it comes to AI development considering highly cited AI patent families, a proxy for commercialization impacts. Nevertheless, this outcome should be questioned because it considers a very long timeframe and private investment in AI R&D in China is a relatively recent phenomenon (Zhu et al., 2018). But even questioning this indicator, China not only lags in the commercialization impact of its patents but, more broadly, in AI economy metrics.

China is ranked 9th considering all the economic metrics of the Stanford University 2020 Global Vibrancy Ranking, which includes skill penetration, an AI hiring index, absolute and relative indicators for AI private investment and the number of AI-founded companies in the country. The US has the greatest number of AI start-ups and the highest levels of private equity and venture capital received by AI start-ups (Castro et al., 2019; Zhang et al., 2021). US-based AI start-ups received over 23.6 billion USD in funding in 2020; China ranked second worldwide with 9.9 billion USD (Zhang et al., 2021).

All in all, China is catching up and even forging ahead in some dimensions regarding AI and the Chinese state role in setting China's NIS has been crucial. The openness of the innovation system remains of major importance for China's catching-up in AI. China's global access to hardware, software and AI talent is crucial for its success. At the same time, openness has become increasingly intertwined with geopolitics. In the next section, we focus on the role of the Chinese tech giants that have emerged during this catching-up process. They play important roles in China's AI catching-up, but they also challenge the Chinese state.

6.3   ON THE CO-EVOLUTION OF NATIONAL AND CORPORATE INNOVATION SYSTEMS

Tech giants harvest both data and knowledge beyond their home countries and they organize global corporate innovation systems. By leading global innovation networks, they challenge the autonomy of states. More broadly, to ascertain access to markets, data and technologies abroad and at home, they operate with some degree of autonomy from the state. Yet, tech giants depend on their home state in maintaining dominant positions in global markets and strategic technologies. The other way around, the state has become dependent on the global success of nationally rooted tech giants, with China as a case in point. Tech giants' operation is crucial both for the overall profits (including financial rents which, as intellectual rents, are another form of value appropriation) they harvest abroad and for their role in developing and applying strategic technology such as AI.

We conceptualize China's catching-up in AI as a process of co-evolution between China's NIS and its tech giants' CIS. Alibaba and Tencent have built their strength in AI not only based on the R&D of Chinese universities and public research organizations but also on privileged access to the Chinese market for digital services and data. The highly developed telecom infrastructure and widespread use of mobile phones together with the digital firewall were prerequisites for the formation of Chinese tech giants. Furthermore, according to Jacobides et al. (2021), the state encouraged its tech companies to drive AI adoption and innovation in China.

Tech giants and the Chinese state work together for the development of AI in China and aiming to fulfil their goal of digital China going global as evidenced in their jointly written "Artificial Intelligence Standardization White Paper" of 2018 and "AI Security Standardization White Paper" of 2019. Alibaba and Tencent have become vanguards in China's ambition to become a world AI leader. The configuration of their CIS confirms that they both give priority to AI and that they harvest knowledge primarily from China—their most frequent publishing collaborators are a handful of Chinese top universities (see Chapters 3 and 4). Nevertheless, in areas where China's NIS is weak, our findings provide evidence of Alibaba and Tencent seeking knowledge and using hardware from abroad.

Moreover, the US and Chinese tech giants' strong positions in their respective domestic markets and their capacity to dictate their own rules and norms in the digital world raise concerns among states. Tech giants' concentration of financial resources and technological competence challenges the sovereign leadership of the state. This becomes especially clear as they develop digital infrastructures encroaching on areas normally under state control (finance, communication, energy and transport).

Just before the Ant Group's IPO in Shanghai, the Chinese government announced antitrust regulations for digital companies forcing Alibaba to cancel the IPO.[15] These announcements were followed by orders to terminate Ant Group's financial activities (credit, insurance and wealth management) keeping only its e-payments business. The state then introduced further requirements to Alibaba, closer to those that commercial banks, which are SOE, must comply with. The same measure will apply to fintech in general, yet Alibaba and Tencent loom large in this market in China (see Chapter 3). In order to remain in control of financial markets, the Chinese state has intervened favouring its SOE over private tech giants. Furthermore, the People's Bank of China accelerated measures to increase the uptake of its digital currency—channelled through those state-owned commercial banks—, hereby establishing a state-controlled competition to AliPay and WeChat Pay. These and recent events outline the ongoing clashes of power between tech giants and the Chinese state, including Xi Jinping's call for stronger regulation of high incomes and common prosperity, which has triggered Alibaba, Tencent and other giants to pledge donations for such aim.

Another point of conflict in the co-evolution of these powerful actors concerns corporate governance, given the significant foreign presence in Alibaba's and Tencent's shareholding structure and the state's recent aims to regulate foreign IPOs.[16] The system that allows foreign capital to own Chinese companies' shares through offshore holding companies, called the Variable Interest Entity (VIE) model, has become a direct channel used by transnational capital to seek control of Chinese digital companies (Shen, 2021; Wójcik & Camilleri, 2015). However, being the main stockowner does not directly translate into controlling Chinese corporations as illustrated by Jack Ma's decision to spin-off Alipay in 2011 as a mainly Chinese owned company to assure the Chinese state's approval to

---

15 https://www.ft.com/content/1a4a5001-6411-45fa-967c-0fd71ba9300b.

16 In line with the globalization of corporate governance (Fichtner & Heemskerk, 2020), by early 2021, Alibaba's top shareholder was the Japanese multinational holding Softbank, with around 25%. Softbank also owns substantial shares in other Chinese tech companies like Didi and ByteDance. Alibaba's second and third-largest stockowners are its co-founders, Joseph Tsai and Jack Ma in that order. After them, leading asset managers from the US and the United Kingdom followed (T. Rowe Price, Vanguard Group, Black-Rock, Baillie Gifford and State Street). Together, the latter owned 10% of the company's shares. The share of Tencent's stocks in foreign hands is even bigger. Although its founder, chairman and CEO, Ma Huateng, held 8% of the company's stocks by that same date, Tencent's largest shareholder is the South African multinational Naspers (31%). After these two major shareholders, three US-based asset managers followed: Vanguard (2%), BlackRock (1%) and Capital Research (1%) (Fernández et al., 2020).

operate in China, regardless of the accusations and discomfort of Yahoo and Softbank, back then Alibaba's main shareholder.[17] The VIE model is another example of the complexities of China's openness strategy. It was established in the early 2000s and the state implicitly agreed on tech company's raising foreign capital ownership (Shen, 2021) until recently, when the China Securities Regulatory Commission was given the task to review Chinese IPO plans abroad that used the VIE model.[18]

All in all, there is often a mutually beneficial relationship between Chinese tech giants and the Chinese state. Compensating for some of the weaknesses of the overall NIS, private tech giants contributes to the realization of state plans to make China a lead country in AI by 2030. But, as illustrated by the recent fintech developments, there are also elements of conflict as tech giants grow bigger and extend their dominant positions to new fields of operation threatening the hegemony of the state. The co-evolution between the national and the corporate systems is, therefore, both conflicting and beneficial for the Chinese state and tech giants.

6.4   Conclusions

China's recent development is an example of how, under specific political and economic circumstances, a transformative technology—in this case AI—opened up a window of opportunity for technological catching-up. In the last decades of the twentieth century, China combined openness to foreign direct investments with major efforts to build a strong domestic knowledge base. The state has played a key role in directing and promoting innovation in strategic technologies. Now the Chinese government sees leadership in AI as a key to catching-up for the whole economy.

China's AI catching-up has been impressive even if it remains second to the US in important areas. A crucial factor underlying China's success is the privileged access to data from the enormous Chinese market. Through the great firewall, Tencent, Alibaba, the other Chinese tech giants and the Chinese state enjoy access to billions of data points originating from multiple aspects of organizations operating in China as well as Chinese

[17] https://www.ft.com/content/40a66dd2-b9ec-11e0-8171-00144feabdc0.

[18] https://www.reuters.com/business/how-chinese-clampdown-will-target-offshore-listings-2021-07-08/.

citizens' lives. The other side of this is the expansion of state surveillance and censorship coupled with growing polarization driven by rising intellectual rents captured by Chinese tech giants.

Will China succeed in its goal to become the world leader in AI in 2030? When Chinese leaders stated this as their explicit goal, they triggered reactions in the rest of the world, especially in the US. The counter-reactions from the US (see the next chapter) and Europe will certainly affect the outcome. To become a world leader, China must build capabilities in the design and production of advanced chips and strengthen its AI fundamental research. This will take time—perhaps more than a decade—and success cannot be taken for granted.

One important handicap for China is its limited access to the most talented scholars in AI at the global level. Here, the market mechanism—offering high economic compensation and attractive material working conditions—may be insufficient, since many leading intellectuals will be reluctant to work with Chinese organizations in the absence of human rights and freedom of speech in China.

We see AI as constituting a critical technology for China's overall catching-up with the high-income countries. The other way around, overcoming current weaknesses in the NIS is fundamental for China's catching-up in AI. While the specific efforts to compensate for weaknesses in hardware, talent and basic research are crucial, equally important is the interaction between knowledge institutions and industry and the innovation capabilities of the majority of enterprises, which ultimately depends on the Chinese state capacity to regulate tech giants and dismantle their intellectual monopolies. These are factors that will impact the capacity to absorb and use AI in new fields such as health, education, intelligent cities, transport and finance.

Further research is needed concerning how the coevolution of national and corporate innovation systems in China and the US shape world development. As another step in this direction, Chapter 7 deals with the US AI strategy and elaborates on the US–China conflict as a technology war building on the distinction between techno-globalism and techno-nationalism.

REFERENCES

Ahmed, N., & Wahed, M. (2020). The de-democratization of AI: Deep learning and the compute divide in artificial intelligence research. *ArXiv Preprint.* ArXiv:2010.15581

Allen, G. C. (2019). *Understanding China's AI strategy: Clues to Chinese strategic thinking on artificial intelligence and national security.* Center for a New American Security.

Azmeh, S., Foster, C., & Echavarri, J. (2020). The international trade regime and the quest for free digital trade. *International Studies Review, 22*(3), 671–692.

Benoliel, D. (2017). *Patent intensity and economic growth* (Vol. 38). Cambridge University Press.

Brynjolfsson, E., & McAfee, A. (2014). *The second machine age: Work, progress, and prosperity in a time of brilliant technologies.* W. W. Norton & Company.

Buckley, P. J., Clegg, L. J., Voss, H., Cross, A. R., Liu, X., & Zheng, P. (2018). A retrospective and agenda for future research on Chinese outward foreign direct investment. *Journal of International Business Studies, 49*(1), 4–23.

Castro, D., & McLaughlin, M. (2021). *Who is winning the AI race: China, the EU, or the United States? 2021 update.* Center for Data Innovation.

Castro, D., McLaughlin, M., & Chivot, E. (2019). *Who is winning the AI race: China, the EU or the United States?* Center for Data Innovation.

China Clarivate Analytics. (2018). *2018 top 100 CHINESE innovators.* Clarivate Analytics.

China Institute for Science and Technology Policy at Tsinghua University. (2018). *China AI development report.*

Colvin, T. J., Liu, I., Babou, T. F., & Wong, G. J. (2020). *A brief examination of Chinese Government expenditures on Artificial Intelligence R and D.* Institute for Defense Analyses Alexandria va.

Ernst, D. (2016). China's bold strategy for semiconductors–Zero-sum game or catalyst for cooperation? *East-West Center Working Papers: Innovation and Economic Growth Series, 9.*

Ernst, D. (2020). *Competing in Artificial Intelligence chips: China's challenge amid technology war.* Centre for International Governance Innovation.

Eun, J.-H., Lee, K., & Wu, G. (2006). Explaining the "University-run enterprises" in China: A theoretical framework for university–industry relationship in developing countries and its application to China. *Research Policy, 35*(9), 1329–1346.

Feng, K. (2019). *Innovation and industrial development in China: A Schumpeterian perspective on China's economic transformation.* Routledge.

Fernández, R., Adriaans, I., Klinge, T. J., & Hendrikse, R. (2020). *The financialisation of Big Tech.* SOMO (Stichting Onderzoek Multinationale Ondernemingen).

Fichtner, J., & Heemskerk, E. M. (2020). The new permanent universal owners: Index funds, patient capital, and the distinction between feeble and forceful stewardship. *Economy and Society*, 1–23.

Freeman, C. (1982). Technological infrastructure and international competitiveness. *Industrial and Corporate Change*, *13*(3), 541–569.

Freeman, C. (1987). *Technology policy and economic performance: Lessons from Japan*. Pinter Publishers. http://tocs.ulb.tu-darmstadt.de/122470664.pdf

Freeman, C. (2007). The political economy of the long wave. In G. Hodgson (Ed.), *The evolution of economic institutions* (pp. 75–97). Edward Elgar.

Fu, X., Sun, Z., & Ghauri, P. N. (2018). Reverse knowledge acquisition in emerging market MNEs: The experiences of Huawei and ZTE. *Journal of Business Research*, *93*, 202–215.

Fu, X., Woo, W. T., & Hou, J. (2016). Technological innovation policy in China: The lessons, and the necessary changes ahead. *Economic Change and Restructuring*, *49*(2–3), 139–157.

Godinho, M. M., & Ferreira, V. (2012). Analyzing the evidence of an IPR take-off in China and India. *Research Policy*, *41*(3), 499–511.

Gu, S., & Lundvall, B. A. (2006a). China's innovation system and the move toward harmonious growth and endogenous innovation. *The Learning Economy and the Economics of Hope*, 269.

Gu, S., & Lundvall, B. A. (2006b). Policy Learning as a key process in the transformation of the Chinese innovation systems. In B. A. Lundvall, P. Intarakumnerd, & J. Vang (Eds.), *Asia's innovation systems in transition* (pp. 293–311). Edward Elgar.

Gu, S., Schwaag Serger, S., & Lundvall, B. A. (2016). China's innovation system: Ten years on. *Innovation*, *18*(4), 441–448.

Guo, L., Zhang, M. Y., Dodgson, M., & Gann, D. (2019). Huawei's catch-up in the global telecommunication industry: Innovation capability and transition to leadership. *Technology Analysis & Strategic Management*, *31*(12), 1395–1411.

Hawes, C., & Chew, E. (2011). The cultural transformation of large Chinese enterprises into internationally competitive corporations: Case studies of Haier and Huawei. *Journal of Chinese Economic and Business Studies*, *9*(1), 67–83.

Hong, Y. (2015). Colonial legacies and peripheral strategies: Social-spatial logic of China's communications development since 1840. *Global Media and Communication*, *11*(2), 89–102.

Humphrey, J., Ding, K., Fujita, M., Hioki, S., & Kimura, K. (2018). Platforms, innovation and capability development in the Chinese domestic market. *The European Journal of Development Research*, *30*(3), 408–423.

International Datacenter Corporation. (2019). *The China datasphere: Primed to be the largest datasphere by 2025*. International Datacenter Corporation.

Jacobides, M. G., Brusoni, S., & Candelon, F. (2021). The evolutionary dynamics of the artificial intelligence ecosystem. *Strategy Science*.

Lee, J., & Lee, K. (2021). Is the fourth industrial revolution a continuation of the third industrial revolution or something new under the sun? Analyzing technological regimes using US patent data. *Industrial & Corporate Change, 30*(1), 137–159. https://doi.org/10.1093/icc/dtaa059

Lee, K. (2013). *Schumpeterian analysis of economic catch-up: Knowledge, path-creation, and the middle-income trap.* Cambridge University Press.

Lee, K., Gao, X., & Li, X. (2017). Industrial catch-up in China: A sectoral systems of innovation perspective. *Cambridge Journal of Regions, Economy and Society, 10*(1), 59–76.

Li, D., Capone, G., & Malerba, F. (2019). The long march to catch-up: A history-friendly model of China's mobile communications industry. *Research Policy, 48*(3), 649–664. https://doi.org/10.1016/j.respol.2018.10.019

Li, R., & Cheong, K.-C. (2017). Huawei and ZTE in Malaysia: The localisation of Chinese transnational enterprises. *Journal of Contemporary Asia, 47*(5), 752–773.

Liu, X., Schwaag Serger, S., Tagscherer, U., & Chang, A. Y. (2017). Beyond catch-up—Can a new innovation policy help China overcome the middle income trap? *Science and Public Policy, 44*(5), 656–669.

Liu, X., Wang, X., & Hu, Y. (2021). *Catch-up and radical innovation in Chinese state-owned enterprises: Exploring large infrastructure projects.* Edward Elgar.

Mueller, M. L. (2011). China and global Internet governance: A tiger by the tail. In *Access contested: Security, identity, and resistance in Asian cyberspace* (pp. 177–194).

National Science Foundation. (2020). *2020 The State of U.S. Science & Engineering.* National Science Foundation.

Nuvolari, A. (2019). Understanding successive industrial revolutions: A "development block" approach. *Environmental Innovation and Societal Transitions, 32,* 33–44.

Pérez, C., & Soete, L. (1988). Catching up in technology: Entry barriers and windows of opportunity. In G. Dosi, C. Freeman, R. R. Nelson, G. Silverberg, & L. Soete (Eds.), *Technical change and economic theory* (pp. 458–479). Francis Pinter.

PWC. (2020). *Global Top 100 companies by market capitalisation.* PWC.

Qi, H., & Kotz, D. M. (2020). The impact of state-owned enterprises on China's economic growth. *Review of Radical Political Economics, 52*(1), 96–114.

Rikap, C. (2021). *Capitalism.* Routledge.

Rikap, C., & Harari-Kermadec, H. (2019). The direct subordination of universities to the accumulation of capital. *Capital & Class.* https://doi.org/10.1177/0309816819852761

Schaaper, M. (2009). *Measuring China's innovation system: National specificities and international comparisons.*

Scherngell, T., Rohde, C., & Neuländtner, M. (2020). The dynamics of global R&D collaboration networks in ICT: Does China catch up with the US? *Plos One, 15*(9), e0237864.

Schwab, K. (2016). *The global competitiveness report 2016–2017: Insight report.*

Shen, H. (2021). *Alibaba: Infrastructuring global China.* Routledge.

Shi-Kupfer, K., & Ohlberg, M. (2019). China's digital rise: Challenges for Europe. *MERICS Papers on China, 7*, 14.

Stanford, H. A. I. (2021). *Artificial intelligence index report 2021.* Stanford University Human Centered Artificial Intelligence.

Tseng, C.-Y. (2009). Technological innovation in the BRIC economies. *Research-Technology Management, 52*(2), 29–35.

Tsui, L. (2007). An inadequate metaphor: The Great Firewall and Chinese internet censorship. *Global Dialogue, 9*(1/2), 60.

UNCTAD. (2019). *Digital economy report 2019: Value creation and capture–Implications for developing countries.* United Nations.

Wen, Y. (2017). *The rise of Chinese transnational ICT corporations: The case of Huawei* [PhD Thesis]. Communication, Art & Technology: School of Communication.

Wójcik, D., & Camilleri, J. (2015). Capitalist tools in socialist hands'? China Mobile in global financial networks. *Transactions of the Institute of British Geographers 40*, 464–478.

World Bank & DRC State Council, The People's Republic of China. (2019). *Innovate China.* World Bank Publications.

World Intellectual Property Organization. (2019). *WIPO technology trends 2019: Artificial intelligence.* WIPO.

Wu, X., & Gereffi, G. (2018). Amazon and Alibaba: Internet governance, business models, and internationalization strategies. In *International business in the information and digital age* (pp. 327–356). Emerald Publishing Limited.

Zhang, C. (2019). *How much do state-owned enterprises contribute to China's GDP and employment?* World Bank.

Zhang, D., Mishra, S., Brynjolfsson, E., Etchemendy, J., Ganguli, D., Grosz, B., Lyons, T., Manyika, J., Niebles, J. C., & Sellitto, M. (2021). The AI index 2021 annual report. *ArXiv Preprint.* ArXiv:2103.06312

Zhou, E. Y., & Stembridge, B. (2008). *Patented in China: The present and future state of innovation in China.* Scientific Thomson Reuters.

Zhu, J., Huang, T., Chen, W., & Gao, W. (2018). The future of artificial intelligence in China. *Communications of the ACM, 61*(11), 44–45.

CHAPTER 7

AI Policies and Politics in China and the US Between Techno-Globalism and Techno-Nationalism

Abstract This chapter focuses on the role of the state and its interaction with tech giants in the US–China race for AI global dominance. In the previous chapter, we elaborated on the role of the Chinese state in China's AI catching-up. In response, the US government has combined measures to *promote* AI with *technology embargos* with the aim to *protect* its AI lead. We use the concepts techno-globalism and techno-nationalism to capture fundamental changes in state involvement and international relations. Simultaneously, the chapter elaborates on the co-evolution between state policy and corporate strategy, characterized by elements of consonance and discord. Tech giants operate globally and their interplay with their home states is at the core of the current transformations of the global innovation system.

Keywords Artificial intelligence · State · Techno-nationalism · US · China · Techno-globalism

7.1 INTRODUCTION

This chapter focuses on the role of the state and its interaction with tech giants in the context of the US versus China race for global dominance in artificial intelligence (AI). While the state in both countries

---

C. Rikap and B.-Å Lundvall, *The Digital Innovation Race*,
https://doi.org/10.1007/978-3-030-89443-6_7

makes use of a series of standard policy instruments to promote science, technology and innovation (STI) (for a recent overview of STI-policy see Chaminade & Lundvall, 2019), the unique characteristics of AI and the fact that both countries define the technology as being of strategic importance is reflected in a wider set of state actions. In both countries, the state supports AI R&D in the public and private sector as well as offering subsidized access to venture capital. Governments invest heavily in digital infrastructure and additional funding is allocated to training new generations of AI experts.

But recent state actions in the US go beyond promoting the development, diffusion and use of AI and complementary technologies. Trade policy is used to protect specific industries and technologies. State support for innovation is directed exclusively towards domestic firms and exclude enterprises with foreign ownership. The US state has established embargos on technologies and denied specific enterprises from China access to the domestic market and it puts pressure on other countries to do the same. Moreover, there is a tendency to undermine the principle of open science as the state makes attempts to block international scientific collaboration with scholars from China.

While the first set of policies can be grounded in neoclassical economic theory as a response to *market failures*, the second set is motivated by references to the establishment of *a level playing field*. The US state presents them as adequate responses to China's unfair competition—disrespect for intellectual property, state subsidies and protectionism—reinforced by a generalized criticism of its political system. The Great Firewall—that makes it difficult for US tech giants to harvest data from China—is seen as especially provoking given the importance of data harvesting in the AI race.

As both countries' states increasingly move towards new and more extreme forms of techno-nationalism, tech giants operate globally and include organizations from the contender country in their corporate innovation systems. For instance, among Microsoft's and Amazon's most frequent collaborations in AI-science we find Chinese universities while Tencent and Alibaba, with strong elements of foreign ownership, locate much of their lead AI research in US hotspots. But, in spite of global operations, tech giants remain either Chinese or US *citizens*. They depend on state's backing and their autonomy vis-à-vis the state is under constant negotiation. Thus, while their main strategy so far has been techno-globalist, when it comes to policy recommendations, they become increasingly "techno-nationalist".

Overall, a general message in this book is that the global development and use of AI need to be understood in the light of an interplay between the state and tech giants—an interplay that reflects both harmony and conflict. In this chapter, we show that tech giants play an important role in defining the current mix of techno-nationalism and techno-globalism. They are shaped and supported by state policies and they serve their own state in some respects, while in others they pursue conflicting aims with the political clout to overtake states. One implication is that it is seriously misleading to design AI national policies and measure AI performance without considering both the citizenship status and the global outreach of US and Chinese tech giants.

In the next section, we briefly compare the discourse on AI as a strategic technology in the US and China. Section 7.3 looks into the US state efforts to defend its technological lead in AI in the context of China's AI catching-up (see Chapter 6). Section 7.4 discusses new trends in state policy through the lenses of techno-globalism and techno-nationalism, and Sect. 7.5 concludes.

7.2   AI IS SEEN AS A STRATEGIC TECHNOLOGY IN CHINA AND THE US

The arguments for defining AI as a strategic technology are quite similar in the two countries.

The US National Artificial Intelligence Research and Development Strategic Plan from 2016 defines AI as "a transformative technology that holds promise for tremendous societal and economic benefit. AI has the potential to revolutionize how we live, work, learn, discover, and communicate. AI research can further our national priorities, including increased economic prosperity, improved educational opportunities and quality of life, and enhanced national and homeland security". A parallel report by the President of the Council of Economic Advisors emphasizes AI's potential impact on productivity and growth.

Similarly, the New Generation Artificial Intelligence Development Plan from 2017 outlines China's AI strategy until 2030, the time China has set to become the world-leading nation in AI. It assumes that AI has great potential to revive the economy, to help reorganize society so that citizens get better social services and to strengthen the military.

While there are few references to the Chinese tech giants, one aim of the Chinese AI plan is to create *world-leading AI backbone enterprises*. In general, the plan assigns an important role to firms and to market regulation. Interestingly, there is not a single reference to the distinction between "private" and "state-owned" enterprises in the document. Chinese AI plan seems to build upon insights from research on innovation, including the literature on catching-up and windows of opportunity in connection with shifts in techno-economic paradigms.

The explicit aspiration to international leadership, together with evidence of China's advancements in AI (see Chapter 6), has played a key role in triggering the US technology war with China. It started with technology embargos directed against specific Chinese tech giants such as Huawei and ZTE. In October 2020, the White House went further and defined 20 specific technologies as being in need of both promotion and protection by the state. Among others, they included AI; energy; quantum information science; communications and networking, including 5G; semiconductors; and space technologies. In March 2021, The US Federal Communications Commission (FCC) designated five Chinese tech firms as posing an "unacceptable risk" to national security. The companies include Huawei, ZTE, Hytera Communications, Hangzhou Hikvision Digital Technology and Dahua Technology.

More recently, the US state has tried to block international scientific cooperation deemed to involve elements of knowledge transfer that could contribute to China's progress in strategic technologies. Specifically, the US government has targeted China's "1000 talents program", which recruits top scholars doing research abroad, especially scholars of Chinese origin. In November 2019, the US Senate Permanent Subcommittee on Investigations and Committee on Homeland Security and Governmental Affairs held an open hearing on China's Talent Recruitment Plans and called the programs a threat to national security. These actions are part of a broader US strategy aimed at limiting China's catching-up and reassuring its AI leadership.

7.3   The US AI Strategy: A Lead Country Defending Its Economic and Military Dominance

The current conflicts between the US and China stand in sharp contrast to the situation just a decade ago when technological and scientific international collaboration was still seen advantageous for both economies

(Xue & Simon, 2013). As long as China was significantly behind in critical technologies, the predominant exchange pattern, where US multinationals got access to the Chinese market while Chinese enterprises attempted to get access to US technology, was uncontroversial.

Yet, as the technological rise of China became apparent, the full power of the US state has been mobilized. It spans from massive state support to build infrastructure and promote innovation to embargos of technologies assumed to be of crucial importance for China's further AI development. If leading the first ICT phase involved winning the technological race against Japan, being at the forefront of this second phase, structured around AI, goes hand in hand with the confrontation with China.

US policy documents recognize the US as the global leader in AI research, development and adoption (The White House, 2020). This statement is in line with several reports that compare the state of AI at the country level (Castro et al., 2019; Castro & McLaughlin, 2021; China Institute for Science and Technology Policy at Tsinghua University, 2018; Stanford, 2021). The US leads from afar in AI start-ups, venture capital and private equity funding, average research quality, designing chips for AI systems and R&D spending by software and computer services firms in the top 2500 firms (Castro & McLaughlin, 2021). However, China is reducing the gap or extending its lead over the US in several metrics (see Chapter 6).

The US AI strategy includes several programs aiming at securing its leadership. In 2017, a new National Security Strategy was launched (The White House, 2017). It calls on the US to lead in research, technology, invention and innovation in emerging technologies, including AI. The centrality of AI for military applications, in particular concerning machine learning, was stated in the US National Defense Strategy.[1] Another sign of the importance given to advanced AI R&D is the *FY 2020 Administration Research and Development Budget Priorities*[2] memo, which calls for investment in AI and quantum computing as critical components of the US national security.

Beyond these initiatives, the overall US strategy for promoting American leadership in AI, called "American AI Initiative", was launched in

---

[1] https://dod.defense.gov/Portals/1/Documents/pubs/2018-National-Defense-Strategy-Summary.pdf.

[2] https://trumpwhitehouse.archives.gov/wp-content/uploads/2018/07/M-18-22.pdf.

2019. It identified five key working lines: increasing—later on rephrased as doubling by fiscal year 2022—AI research investment, expanding Federal AI computing and data resources, setting AI technical standards, building America's AI workforce and engaging with international allies. This was followed by specific initiatives at the level of each federal agency (the Department of Defence, the NIH, DARPA, the USPTO, etc.).[3] In the fiscal year 2020, the US unclassified the information that non-defence annual Federal investment in AI R&D was 973.5 million USD. It was the first time this figure was calculated (The White House, 2020).

The National AI R&D Strategic Plan update of 2019 defined eight lines of action: (1) Make Long-Term Investments in AI Research, (2) Develop Effective Methods for Human-AI Collaboration, (3) Understand and Address the Ethical, Legal and Societal Implications of AI, (4) Ensure the Safety and Security of AI Systems, (5) Develop Shared Public Datasets and Environments for AI Training and Testing, (6) Measure and Evaluate AI Technologies through Standards and Benchmarks, (7) Better Understand the National AI R&D Workforce Needs, and (8) Expand Public–Private Partnerships to Accelerate Advances in AI.

The US does not explicitly refer to China in those policy documents but includes other countries as international allies—such as the UK and France—and as strategic to the development of its AI leadership. Furthermore, it does mention authoritarian governments or authoritarian uses of AI as threats to be countered through reassuring the US AI leadership.

> The best way to counter authoritarian uses of AI is to make sure America and our international partners remain the global hubs of innovation, shaping the evolution of technology in a manner consistent with our common values. These AI regulatory principles put our Nation on a path towards continued AI leadership, innovation, and discovery. (The White House, 2020)

In contrast, a report commissioned to an ad-hoc group of advisors, the National Security Commission on Artificial Intelligence (NSCAI), is extremely vocal on China as an adversary. The report continuously refers to China's AI threat and the need to dedicate greater efforts to confront China's ambition to lead AI in 2030.

---

3 https://trumpwhitehouse.archives.gov/ai/executive-order-ai/.

For the first time since World War II, America's technological predominance—the backbone of its economic and military power—is under threat. China possesses the might, talent, and ambition to surpass the United States as the world's leader in AI in the next decade if current trends do not change. Simultaneously, AI is deepening the threat posed by cyber attacks and disinformation campaigns that Russia, China, and others are using to infiltrate our society, steal our data, and interfere in our democracy. (National Security Commission on Artificial Intelligence, 2021)

The NSCAI illustrates the close cooperation between tech giants and the state. It was chaired by Eric Schmidt, Google's former chairman, and included the head of Google Cloud Artificial Intelligence division, Microsoft's Chief Scientific Officer and Andrew Jassy, founder and former CEO of Amazon Web Services (AWS) and, since July 5, 2021, Amazon's CEO. Another example of cooperation was the National Science Foundation's Program on Fairness in Artificial Intelligence which was funded by a joint grant with Amazon.[4] Moreover, Eric Schmidt and Bill Gates were invited by the New York Governor to *re-imagine* education after the Covid-19 pandemic by applying new technologies.[5]

The NSCAI's main thesis is that the US is not sufficiently prepared to defend its lead in the AI era. To be AI-ready, the report argues, the US must overcome its vulnerabilities on three fronts: data, computing power and talent (National Security Commission on Artificial Intelligence, 2021). The document gives recommendations to tackle each of these issues, from establishing a Digital Service Academy to revitalize domestic microchip fabrication to become less dependent on TSMC (among others because of its factories are mainly located in China).

Chips are of paramount importance both for the US and China, but—as we mentioned in Chapter 6—the US currently has an advantage. Taiwan's TSMC and South Korea's Samsung are the world leaders in production capacity, with a clear dominance of TSMC. They are the only companies that produce 5nm chips, which have the smallest transistors

---

[4] https://www.nsf.gov/funding/pgm_summ.jsp?pims_id=505651 last access May 27, 2021.

[5] https://www.washingtonpost.com/education/2020/05/06/cuomo-questions-why-school-buildings-still-exist-says-new-york-will-work-with-bill-gates-reimagine-education/ last access February 20, 2021.

therefore the lowest energy consumption and highest speed. In 2020, 60% of TSMC sales were exported to the US and only 20% to China.[6]

As we mentioned in Chapter 6, it is impossible to produce advanced semiconductors without access to leading US players' design technology.[7] This dependence gives the US veto power over chip producers' sales to China—they were for instance forced to cut off supplies to Huawei and ZTE in 2020. Its strength was also used to pressure chip producers to establish manufacturing facilities in the US.[8] In 2020, the US government agreed with TSMC the opening of a plant in Phoenix. The US government wants TSMC to produce chips for the US military, such as those used in F-35 stealth fighter jets, on US soil.[9]

Overall, the NSCAI report makes a strong case for techno-nationalism, stimulating the collaborations between US companies and federal agencies but simultaneously stating that most AI R&D, while financed by the state, should be done by private firms and universities. They suggest combining the US garage start-up mentality with massive US state investments. Explicitly referring to tech giants' capabilities in the tech war with China, the report claims that "even large tech firms cannot be expected to compete with the resources of China or make the big investments the U.S. will need to stay ahead. We will need a hybrid approach meshing government and private-sector efforts to win the technology competition" (National Security Commission on Artificial Intelligence, 2021, p. 25). This hybrid approach is quite beneficial for tech giants because they engage the US government in voluminous R&D investments while keeping the means to collect associated gains.

In line with these claims and also with the diagnosis of the US weakness, in 2021 Biden launched a 2 trillion USD infrastructure plan called "The American Jobs Plan". It includes 180 billion USD for R&D in AI and biotechnology, aimed at overcoming the Chinese threat. Another

---

6 https://www.ft.com/content/b452221a-5a82-4f5d-9687-093b9707e261?segmen tId=b0d7e653-3467-12ab-c0f0-77e4424cdb4c.

7 https://asia.nikkei.com/Spotlight/The-Big-Story/US-China-tech-war-Beijing-s-sec ret-chipmaking-champions.

8 https://www.ft.com/content/b452221a-5a82-4f5d-9687-093b9707e261?segmen tId=b0d7e653-3467-12ab-c0f0-77e4424cdb4c and https://en.yna.co.kr/view/AEN202 10517005400320.

9 https://www.ft.com/content/b452221a-5a82-4f5d-9687-093b9707e261?segmen tId=b0d7e653-3467-12ab-c0f0-77e4424cdb4c.

300 billion USD will be devoted to manufacturing subsidies, including major subsidies for chipmakers. The Biden administration recognized the strategic centrality of chips and, therefore, the need to invest in chip infrastructure inside the US to overcome shortages and reassure the US technological leadership. The NSCAI report concludes along the same lines and goes even further suggesting that "rebuilding domestic chip manufacturing will be expensive, but the time to act is now. The United States should commit to a strategy to stay at least two generations ahead of China in state-of-the-art microelectronics and commit the funding and incentives to maintain multiple sources of cutting-edge microelectronics fabrication in the United States" (National Security Commission on Artificial Intelligence, 2021, p. 12).

Finally, the NSCAI report calls for more stringent intellectual property rights (IPRs) stating that "U.S. courts have severely restricted what types of computer-implemented and biotech-related inventions can be protected under U.S. patent law. Critical AI and biotech-related inventions have been denied patent protection since 2010" (National Security Commission on Artificial Intelligence, 2021, p. 201). The report argues that this has led inventors to prefer trade secrecy, which in turn stifles innovation at the system level. It, thus, advocates for more stringent IPRs, not only for AI innovation but also for data, to reassure the US IPRs policy leads vis-à-vis China and limits what they claim is the latter's steal of US technology. Again, this policy directly favours tech giants while limiting opportunities for smaller and new players.

Summing up, the US state is active in keeping its AI leadership and, in this strategy, US tech giants have a privileged advisory role, thus a crucial place in policymaking in aspects that are at the basis of their intellectual monopolies. The interdependency between the state and the corporations is also apparent in the case of surveillance (see Box 7.1). In the next section, we assess these and other US (and Chinese) policies through the lenses of techno-nationalism and techno-globalism.

Box 7.1 Surveillance: A partnership between the US state and tech giants
Tech giants not only influence the US overall AI policy. Surveillance enabled by AI is a specific field where US big tech companies' business and

the US state requirements come together. Tech giants have been instrumental in the US state surveillance, a neglected phenomenon as compared to the link between the Chinese state's surveillance and its tech giants. The most evident example is that Google, Apple, Facebook and other US internet companies have provided data access to the US National Security Agency (NSA) for global surveillance.

Centralized data of world citizens and organizations is a paramount source of political power that can be used in exchange for economic power or to resist government regulations. In particular, high-tech data-driven intellectual monopolies have been behind web data policies: in their platforms they define data security and data privacy policies and are also entitled to decide whether to ban content. Hence, they have global political power through their control of digital circulation of information.

AI technologies provided by tech giants to the US government span from facial recognition for police departments to the Joint Enterprise Defense Infrastructure (JEDI).

Microsoft and the NYPD have worked together to create a platform called Domain Awareness System since 2009. "By 2016, the system had ingested 2 billion license plate images from ALPR cameras (3 million reads per day, archived for five years), 15 million complaints, more than 33 billion public records, over 9,000 NYPD and privately operated camera feeds, videos from 20,000-plus body cameras and more. To make sense of it all, analytics algorithms pick out relevant data, including for predictive policing" (Kwet, 2020).

JEDI was a large US Department of Defense cloud computing contract, which has been reported worth 10 billion USD. It was initially given to Microsoft, and it is now subjected to a new bid where Amazon and Microsoft are expected to share the deal. Amazon had already developed a 600 million USD computing cloud for the Central Intelligence Agency in 2014 and Google had sizeable contracts with the Pentagon. Another recent example is the contract that Microsoft won in 2021, worth 21.9 billion USD, to provide the US army with 120,000 augmented reality headsets. Its "Integrated Visual Augmentation System" headsets will help soldiers train, home in on targets or be aware of nearby threats by overlaying contextual information on top of the real world.

Zuboff (2015) has shown how, under surveillance capitalism, corporations accumulate and control markets by predicting and modifying human behaviours. To this we need to add their role in assisting surveillance by the state at home and abroad. The surveillance potential of some of the technologies currently being developed by US tech giants—such as facial recognition and video surveillance—is of paramount value for governments. Furthermore, they raise serious concerns regarding human rights and privacy. The state-corporate use of these technologies fosters closer relationships between the US state and its data-driven intellectual monopolies.

7.4   A World Economy Between Techno-Globalism and Techno-Nationalism

While there is no example of extreme techno-globalism or techno-nationalism, these concepts help us understand emerging tendencies and countertendencies in the world system of innovation, in particular in relation to AI as a strategic technology.

The old neoclassical economics growth models were based on a strictly techno-globalist vision since it was assumed that technology was free to access for all, independent of location and nationality. Reality has never been close to such a world. Technology comprises both information that can float freely and sticky knowledge in the form of tacit and experience-based know-how. At the core of successful catching-up processes, we find active governments engaged in building stronger national innovation systems and protecting infant industries. Some have also aimed at reducing intellectual rents paid to multinationals and dependence on foreign knowledge. In response, the US and other lead countries, have reacted by condemning what they refer to as unfair *techno-nationalism* in the upcoming countries.

Nelson and Ostry (1993) is the classical reference for distinguishing between techno-globalism and techno-nationalism. It was part of a series of studies initiated by the Brookings Institute on International Relations and Consequences of Globalisation and it was based on the idea that the US hegemony was coming to an end—at that time the rise of Japan was seen as the main challenge for the US. Nelson and Ostry (1993) defined techno-nationalism as protectionist state interventions aiming at promoting high-technology industries. The difficulty for US multinationals to establish themselves in Japan through foreign direct investments and the European support to Airbus were presented as major examples of techno-nationalism provoking US countermeasures. However, the authors also pointed out that the US state had given more subsidies than most other countries to the private sector's development of science and technology, including Japan. Still, the subsidies appeared under headings of health and security policy, not under the label of industrial policy.

Montresor (2001) introduced three crucial dimensions for understanding techno-nationalism. A country may be regarded as a *territory*— seen from this perspective all technological capabilities located within the national borders may be seen as constituting a national technological system. However, the state may have more or less control of technological development within its territory reflecting the degree of technological *sovereignty*. This is related to the third dimension, the question of *citizenship*: to what degree are technology-intensive foreign companies operating in the country and domestic multinational corporations under the control of the state? Or, are they primarily to be seen as agents for a foreign state? These issues take on a new actuality in the era of digital technologies where some of the tech giants are rulers of global digital platforms. In the concluding section, we will argue that for third countries the issues of technological sovereignty and of the citizenship of the tech giants that they host, directly impact on those countries' development opportunities.

Government initiatives to catch-up in strategic technologies, use of embargos to hinder followers to catch up and efforts to attract specialized skilled workers from abroad are recurring phenomena in history. The very first examples go back to the sixteenth century when England developed ironworks to produce cannons for its fleet and more recent ones include the US Manhattan project (Chaminade & Lundvall, 2019, p. 2). Nonetheless, the technological rise of Japan in ICT—based on an active government, reverse engineering and the formation of Keiretsus that limited market access for US multinational firms—was seen by the US state as a challenge to its dominance in microelectronics and triggered sanctions aiming at slowing down the Japanese advance (Freeman, 1987).

Although a certain element of techno-nationalism has proven to be instrumental for technological catching-up, there are no examples in modern history of full techno-nationalism, where domestic technologies are based exclusively on knowledge produced within a country. Historically, universities were born open to knowledge sharing (de Ridder-Symoens, 1992) and even when they turned towards more national institutions, exchanges between intellectuals did not cease (Le Goff, 1985). The flow of experts and intellectuals travelling across borders and bringing with them technical know-how as well as scientific knowledge has also contributed to international knowledge diffusion. Furthermore, while technologies can be made artificially scarce through IPRs or because they are based on tacit knowledge and secrecy, learning by reverse engineering or through licensing the technology remained possible.

There is little doubt that the porousness of national borders, allowing knowledge to flow between countries, has been one important factor behind major technological developments. However, this relative openness has not served all countries equally well. Companies in lead countries have accumulated IPRs and their states have joined forces to impose rules for latecomer countries that limit their freedom to engage in catching-up strategies. Since the mid-1990s, a stringent international IPRs regime was established through the TRIPS agreement and other bilateral trade agreements that limited access to knowledge as well as reverse engineering learning (Drahos, 1995; Dreyfuss & Frankel, 2014).

Nelson and Ostry (1993) saw the US transnational companies as moving the world towards techno-globalism. Tech giants may be a case in point. With a strong focus on IPRs and tech giants harvesting knowledge and data from abroad, the US imports knowledge and data as well as intellectual rents. Hosting the enterprise vanguards of both the first and the second phase of the ICT revolution has cemented its role as the world's technology leader. This is fundamental for its geopolitical hegemony since its control of the dollar as world currency and its worldwide military presence in part rely on its technological lead (Schwartz, 2019). Therefore, it is not surprising that the US, when confronted with a serious contender in a strategic technology, steps back from the techno-globalism implicit in the Washington Consensus and engages in radical forms of techno-nationalism.

As we demonstrated in earlier chapters, tech giants are specialized in digital services and, through their privileged access to data, they established themselves as AI world leaders. While they establish innovation networks with scholars, knowledge institutions and enterprises around the world, they do so to strengthen their technological dominance. Through their harvesting of data and knowledge in host countries, they absorb financial wealth as well as technological capabilities that are in part entered to their own state. In these respects, they have citizenship in their home countries.

Proximity to their respective home state has been reinforced by the new forms of state action, where tech giants become increasingly and more directly drawn into the technology war. An example is the US ban on using Huawei networking equipment to block its development of the new 5G telecom technology. Google early on responded by restricting Huawei's access to essential smartphone apps. Gmail, YouTube, Google Drive and even the Google Play Store itself are now no longer available, forcing Huawei to use or develop new apps.[10]

At the same time, the very size and mode of operation of the tech giants in the US and China threaten the sovereignty of the state in those countries. The fact that Facebook could block access for the President of the US to a medium with monopoly status illustrates this problem. Other examples are Facebook's attempts to establish Libra as a global currency and Apple's denial to unlock a Saudi air force officer's iPhone in what the US government considered as security issues (Shubber, 2020). Alibaba and Tencent overtaking part of the business of state-owned commercial banks in China is also illustrative (see Chapter 6). Such developments provoke political reactions and state restrictions of tech giants.

7.4.1   Is Techno-Nationalism Bad?

Techno-nationalism takes on different forms and the consequences for the world economy differ depending on the context, in particular, it makes a difference if is it practised by a lead country or by an emerging or underdeveloped economy.

In a regime of relative techno-globalism, where there is open access to data and knowledge resources, endogenous forces work towards polarization. Lead countries benefit from intellectual rents and their knowledge-intensive sites attract knowledge from the rest of the world (see Chapter 4 on AI and geographical concentration). In such a world, successful catching-up would constitute a major contribution to the creation of global wealth. High rates of economic growth in the first decades of the post-war period were a result of Europe's and Japan's catching-up with the US. More recently, China's catching-up has countered the tendency towards global economic stagnation. When the state of an emerging economy successfully engages in supporting knowledge creation and protecting infant industries of strategic importance for the

---

[10] Retrieved from https://www.wsj.com/amp/articles/u-s-pushing-effort-to-develop-5g-alternative-to-huawei-11580831592 last access August 13, 2021.

overall innovation process and innovation system, it contributes both to global growth and to the reduction of global income equality.

Successful performance in international competition in advanced technologies requires, among others, high-quality research institutions, skilled workers, experienced engineers and managers. It also requires to build infrastructure and a well-functioning financial system which is a quite demanding task and it takes time to do it. One way to capture these elements is to define "national innovation systems" (Freeman, 1987; Lundvall, 1992; Nelson, 1993). Narrow definitions include private and public organizations engaged in science and technology. Broad definitions include institutions and organizations related to education and training, organizational forms promoting learning by doing as well as financial institutions (Chaminade et al., 2018; Lundvall et al., 2009). The system perspective highlights the importance of establishing and reshuffling linkages between producers and users of knowledge (Lundvall, 1988). As we saw for the US (Sect. 7.2) and China (Chapter 6), both states have made use of all these instruments to promote the development, diffusion and use of AI.

Given the specific characteristics of AI, it is obvious that even the broad definition of the innovation system and derived STI policy are too narrow. Issues about sovereignty and power asymmetries are crucial for policy outcomes. For third-party countries, a prerequisite for development is a state with sufficient enforcement capacities to limit tech giants' knowledge extractivism, otherwise undermining the positive impact of domestic investments in knowledge. In this context, it is significant that neither in China nor in the US the tech giants emerged without state support. Both states laid the foundation for AI and for their tech giants through measures that went far beyond "innovation policy".

7.4.2   How the State Laid the Foundation for AI and Tech Giants in China and the US

The state has played a major role all through the history of AI. The very first steps taken by Alan Turing were part of England's war efforts and reflected an attempt to decode German war communication. At the beginning, the main driver was political rather than economic. This is true also for the state's role underlying the emergence of the tech giants in China and the US.

In the US, the state's science and technology strategy fulfilled multiple aims. It was vital to guarantee military superiority, defend and increase the power of US multinational corporations and limit other countries' catching-up (from the USSR during the Cold War to Japan in the 70s and China nowadays) (Block, 2008; Wade, 2017). Technological leadership has been fundamental for US geopolitical hegemony (Weiss, 2014).

Military initiatives in promoting ICT are a central precedent of the US current leading position in AI. DARPA launched the first versions of the internet and the Pentagon successfully engaged university research in developing related technologies. This was combined with a broader long-term strategy that contributed to the emergence of tech giants, such as the extension of IPRs (Coriat & Orsi, 2002; Drahos, 1995) and the weakening of antitrust regulation that took place since the late 1970s (Glick, 2019). Permissive antitrust policy reached a climax during Trump's presidency, with fewer criminal antitrust cases brought to courts than any administration since the 1970s (Shubber, 2019).

One of the arguments put forward to undermine China's technological advance is that it is heavily subsidized by the state—which would imply that China is not respecting the idea of a *level playing field*. When it comes to STI-policy, yet this picture is misleading. According to Băzăvan (2019), the direct intervention in terms of government funding and subsidies for enterprise research is quite modest in China as compared to other countries, including the US—this is also the pattern that comes out of OECD statistics. Selective support to high-technology start-ups and stimulation of experiments and entrepreneurship are more important than R&D subsidies. This does not rule out massive support from the state to narrowly defined specific technology fields—examples could be solar and wind technologies and technologies crucial for the AI field.

Finally, an outstanding reason why China could become a challenger to the US in AI and the prerequisite for the emergence of its tech giants is that the state introduced the Great Firewall and, as we saw in the previous chapter, this reflected mainly the Chinese state's concerns about controlling the traffic on the internet. It was not driven primarily by economic or innovation aims. This was coupled with the development of advanced AI research capabilities at universities and public research organizations which, as presented in Chapter 6, are indispensable for explaining China's AI positioning and the emergence of its tech giants.

7.5   FINAL REMARKS

In response to China's catching-up in AI and the emergence of its tech giants, the US state—influenced by its own tech giants—has combined measures to promote AI with technology embargos aimed at protecting its lead in AI. The US position in relation to China has not changed significantly since Biden replaced Trump as the US president. Although the language and style have been changed—more emphasis on human rights—the underlying logic seems to be the same. This illustrates that there is a clear state policy to thwart other countries' attempts to challenge the US hegemony, with a lead in strategic technologies as central part of the equation.

The US and Chinese states' techno-nationalist actions overviewed in this chapter overlayer the general tendency towards privatization of knowledge. What is taking place may be seen as a *double enclosure process* undermining the global knowledge commons and open science. Tech giants privatize, monopolize and turn important elements of technology into private assets while their respective state builds new barriers to the international flow of knowledge. While this may result in inflated innovation indicators both in China and the US, it curtails innovation possibilities for other organizations and for the rest of the world. A definite consequence is growing income inequality and growing global digital divides between AI producers and users. In the next and concluding chapter, we take a global perspective arguing that in the light of the urgency of global challenges there is a need for new forms of global governance and knowledge sharing.

REFERENCES

Băzăvan, A. (2019). Chinese government's shifting role in the national innovation system. *Technological Forecasting and Social Change, 148*, 119738.

Block, F. (2008). Swimming against the current: The rise of a hidden developmental state in the United States. *Politics & Society, 36*(2), 169–206.

Castro, D., & McLaughlin, M. (2021). *Who is winning the AI race: China, the EU, or the United States? 2021 update.* Center for Data Innovation.

Castro, D., McLaughlin, M., & Chivot, E. (2019). *Who is winning the AI race: China, the EU or the United States.* Center for Data Innovation.

Chaminade, C., & Lundvall, B. A. (2019). Science, technology, and innovation policy: Old patterns and new challenges. *Oxford Research Encyclopedia of Business and Management.*

Chaminade, C., Lundvall, B. A., & Haneef, S. (2018). *Advanced introduction to national innovation systems*. Edward Elgar Publishing.

China Institute for Science and Technology Policy at Tsinghua University. (2018). *China AI development report*.

Coriat, B., & Orsi, F. (2002). Establishing a new intellectual property rights regime in the United States: Origins, content and problems. *Research Policy, 31*(8–9), 1491–1507.

de Ridder-Symoens, H. (1992). *A history of the university in Europe: Volume 1, Universities in the Middle Ages* (Vol. 1). Cambridge University Press.

Drahos, P. (1995). Global property rights in information: The story of TRIPS at the GATT. *Prometheus, 13*(1), 6–19.

Dreyfuss, R., & Frankel, S. (2014). From incentive to commodity to asset: How international law is reconceptualizing intellectual property. *Michigan Journal of International Law, 36*(4), 557–602.

Freeman, C. (1987). *Technology policy and economic performance: Lessons from Japan*. Pinter Publishers. http://tocs.ulb.tu-darmstadt.de/122470664.pdf

Glick, M. (2019). Antitrust and economic history: The historic failure of the Chicago School of antitrust. *The Antitrust Bulletin, 64*(3), 295–340.

Kwet, M. (2020). The microsoft police state: Mass surveillance, facial recognition, and the azure cloud. *The Intercept*. https://theintercept.com/2020/07/14/microsoft-police-state-masssurveillance-facial-recognition/

Le Goff, J. (1985). *Les intellectuels au moyen âge*. du Seuil.

Lundvall, B. A. (1988). Innovation as an interactive process: From user-producer interaction to national systems of innovation. In G. Dosi, C. Freeman, R. Nelson, G. Silverberg, & L. Soete (Eds.), *Technical change and economic theory*. Pinter Publishers.

Lundvall, B. A. (1992). *National systems of innovation: Toward a theory of innovation and interactive learning* (Vol. 2). Pinter Publishers. https://books.google.com.ar/books?hl=es&lr=&id=20qCC6MmYgcC&oi=fnd&pg=PR13&dq=National+systems+of+innovation.+Towards+a+theory+of+innovation+and+interactive+learning&ots=QAlvMXp3O3&sig=J5PPsv2olldI8BKN25snLUtJYJ0

Lundvall, B. A., Vang, J., Joseph, K. J., & Chaminade, C. (2009). Innovation system research and developing countries. In *Handbook on innovation systems and developing countries: Building domestic capabilities in a global setting* (pp. 1–31). Edward Elgar.

Montresor, S. (2001). Techno-globalism, techno-nationalism and technological systems: Organizing the evidence. *Technovation, 21*(7), 399–412.

National Security Commission on Artificial Intelligence. (2021). *Final Report*.

Nelson, R. R. (1993). *National innovation systems: A comparative analysis*. Oxford University Press. https://books.google.com.ar/books?hl=es&lr=&

id=T8sJCAAAQBAJ&oi=fnd&pg=PT7&dq=nelson+robert+1993+innova
tion&ots=YBHC2MB_Az&sig=KnE-ju9UjngHsafQl2hIu4MZiJg

Nelson, R., & Ostry, S. (1993). *Techno-nationalism and techno-globalism: Conflict and cooperation.* The Brookings Institution.

Schwartz, H. M. (2019). American hegemony: Intellectual property rights, dollar centrality, and infrastructural power. *Review of International Political Economy,* 1–30.

Shubber, K. (2019, November 5). US price-fixing prosecutions at historic low for third straight year. *Financial Times.* https://www.ft.com/content/a3b75c80-fe74-11e9-be59-e49b2a136b8d

Shubber, K. (2020). US presses Apple to unlock iPhones of Saudi gunman. *Financial Times.* https://www.ft.com/content/eb3cf780-3642-11ea-a6d3-9a26f8c3cba4

Stanford, H. A. I. (2021). *Artificial Intelligence Index Report 2021.* Stanford University Human Centered Artificial Intelligence.

The White House. (2017). *National Security Strategy of the United States of America.* Presidency of the United States.

The White House. (2020). *American Artificial Intelligence Initiative: Year One Annual Report.* The White House.

Wade, R. H. (2017). The American paradox: Ideology of free markets and the hidden practice of directional thrust. *Cambridge Journal of Economics, 41*(3), 859–880.

Weiss, L. (2014). *America Inc.? Innovation and enterprise in the national security state.* Cornell University Press.

Xue, L., & Simon, D. (2013). US-China science and technology cooperation. *China-US Focus.*

Zuboff, S. (2015). Big other: Surveillance capitalism and the prospects of an information civilization. *Journal of Information Technology, 30*(1), 75–89.

CHAPTER 8

Alternative Futures and What is to Be Done

Abstract In this final chapter, we introduce alternative scenarios concerning tech giants, the US and Chinese states, data governance and innovation under the current governance regime. On this basis, we advance policy recommendations and calls for activism aiming at a less polarized future where technology is guided towards collectively solving societal, ecological and health challenges. We elaborate on why there is a need for new forms of governance that take into account the growing importance of global digital public goods as well as the need for new forms of public access to science and technology. We end by applying a global perspective pointing to major dilemmas between what can be done under existing governance regimes and what should be done in the long term.

Keywords Antitrust · Digital acts · Policy recommendations · Global public goods · Activism

8.1 INTRODUCTION

In this book, we have studied crucial aspects of the innovation race in artificial intelligence (AI) between the hegemon—the US—and its challenger China (Chapter 7), and we have seen how globally oriented tech giants

C. Rikap and B.-Å Lundvall, *The Digital Innovation Race*,
https://doi.org/10.1007/978-3-030-89443-6_8

from these countries play key roles in developing, applying and profiting from this general-purpose technology (Chapters 4 and 5).

The intense and conflictual character of the US–China relationship reflects that AI has the potential to change the world's pecking order. At the corporate level, the efforts to win the race are equally intense, since the use and development of AI is a crucial factor driving the concentration of intangible capital by tech giants. We have conceptualized these companies as data-driven intellectual monopolies (see Chapter 2) and explained how they respectively lead and organize corporate innovation systems (Chapter 3). These systems integrate subordinate organizations, including universities and start-ups. The overall R&D orientations are set by the intellectual monopoly, which therefore garners most of the intellectual rents resulting from transforming data and research into assets.

As we have shown in Chapter 6, the co-evolution of national and corporate innovation systems is complex and full of contradictions. China and the US make efforts to support their own intellectual monopolies while the tech giants try to safeguard a certain degree of autonomy in relation to the state. At the same time, the extreme concentration of financial resources and economic power in tech giants creates social imbalances and challenges the power of the state and this provokes efforts to regulate their activities in the US as well as in China.

The global innovation race affects all regions and countries and shapes the AI technological trajectory in terms of both speed and direction. While the engagement of powerful states and enterprises in technological competition accelerates the rate of technological change in certain dimensions and directions, it does so in an uneven and lop-sided fashion, increasing global inequalities and aggravating global problems. Furthermore, tech giants' intellectual monopolies curtail public access to knowledge thereby reducing the overall global rate of innovation, as reflected in recent discourses on secular stagnation.

The current pattern of development, where the AI trajectory is subsumed under the national interests of two major countries and exploited in the interests of a handful of intellectual monopolies, opens up a huge and growing gap between what could be achieved with this technology and what is actually achieved.

These developments raise a series of new political issues in respectively the US, China, other core regions like Europe and the rest of the world. In this chapter, we explain why current political instruments at the disposal of the single (national or regional) state are insufficient when it comes to responding to these new challenges. AI has the potential to create enormous wealth and could contribute to tackling global challenges such as ecological disasters and pandemics. However, this would require new and more ambitious forms of international cooperation and new forms of governance at the national level.

So far, the focus of state intervention and policy debates has been either on tech giants' market power and tax avoidance or on (data) privacy issues. In this book, we have focused on a third dimension: who has the power to shape the technologies of the future, and who will benefit from them? We see the aggravation of the US–China conflict as rooted in this dimension and it is, of course, a major issue for the rest of the world.

A classical policy response at the national level would be to engage in the innovation race through active science, technology and innovation (STI) policy. It could take the form of increased efforts to promote AI through public and private research, education and establishing links between users and producers of AI. While massive efforts to stimulate innovation in AI through standard STI policies in the US and China contribute to their techno-nationalist agendas, such policies will have limited positive effects in the rest of the world. They are necessary but not sufficient to counter the extreme concentration of AI innovation and associated profits.

In the rest of the world, STI policies aiming at building AI competence will help to spread the use of the technology in different sectors. But, at the same time, they will strengthen tech giants' cloud computing business (see Chapter 4) and fail in establishing domestic corporations as major AI developers and producers. The major reason is the character of AI, which results in tech giants' unique capacity to harvest the fruits of knowledge where others have sown (see Chapters 3–5).

The world is at a pivotal stage where the major alternatives are a return in the direction of a more open exchange of data and technologies and a balkanization of the global knowledge landscape. As far as we can judge, the latter alternative is the more likely. This is worrying in a situation where the pandemic, ecological crises and (income, wealth, gender and racial) inequalities require a new kind of collaboration and common efforts to develop and share scientific results and technological know-how.[1] In a world where the two lead countries engage in techno-nationalist strategies, the rest of the world is forced to reconsider

---

[1] Neither would it be sufficient to move back to the old form of globalization since it fosters and reproduces economic inequality and ecological problems both domestically

its openness in important respects, including data, foreign acquisitions, access to foreign technology programs and scientific collaboration.

In a period characterized by the combination of technological revolution and growing geopolitical conflict, the future outcome of the global innovation race remains uncertain. Therefore, this chapter explores alternative futures focused on who will dominate the shaping of AI trajectory under the current governance regime. We also overview existing policy responses to the emergence of data-driven intellectual monopolies, elaborate on their limitations and explore alternative or complementary policies. Yet, government action is not a sufficient response. At least as important are people's agency and awareness. Citizen actions based upon a deeper understanding of how AI works and how it is being used by tech giants and the most powerful states are required to develop the kind of new institutions and forms of governance that can counterbalance their power.

8.2   Predominant Policy Responses to Tech Giants: Antitrust and Digital Acts

As we mentioned above, most of the policy debate in relation to tech giants and digital capitalism has focused on antitrust and data governance. The US Congress (2020) recently investigated (and subsequently sued) tech giants' excessive market power. And just before Ant Group's IPO in Shanghai, the Chinese government announced antitrust regulations for digital companies forcing the cancellation of the IPO.[2] Yet, the earliest concerns on these topics came from the European Union (EU).

8.2.1   The European Union

The EU was among the first to raise awareness of US tech giants' abusive market power. Since 2010, it has run three antitrust investigations against Google: Google Shopping (2010), Google's Android (2015) and Google AdSense (2016). In 2017, the EU competition agency solved the first claim. It found Google guilty of systematically ranking its websites

---

and internationally. In the final part of this chapter, we point to the need for new forms of international cooperation and national governance.

2 https://www.ft.com/content/1a4a5001-6411-45fa-967c-0fd71ba9300b.

and services first, thus favouring its own businesses vis-à-vis competitors. Google was fined €2.42 billion (European Commission, 2017). Concerning Google's Android case, in 2018 the agency "fined Google €4.34 billion for illegal practices regarding Android mobile devices to strengthen dominance of Google's search engine".[3] The European Commission also ruled against Apple and Ireland in 2016 finding that the latter granted illegal state aid to the former through selective tax breaks. Yet, the EU General Court annulled this decision in 2020. The dispute is still unsettled.

Since the Covid-19 pandemic, European countries' antitrust concerns regarding US tech giants have increased. For instance, antitrust regulators in the UK, Germany and Australia are jointly investigating the domination of internet giants.[4] Germany had already blocked Facebook from merging data from its own services in 2019.[5]

Europe's active stand reflects that—this time—it risks ending on the subordinate side, where the peripheries have historically been and generally remain. This is reflected in aggregate data on the location of the world's biggest companies. With a drop of 8 companies between March 2009 and December 2019, Europe's share of global top 100 corporations in market capitalization fell from 27 to 15%. This drop was corresponded by a growing share for the US (PWC, 2020). Regulating the digital economy could thus be seen as Europe's geopolitical rebalancing move.[6]

Antitrust regulations have been complemented with Europe's general data protection regulation (GDPR) and, recently, with a proposed Artificial Intelligence Act. More broadly, the European Commission has made it explicit in recent reports that it is aware of and concerned about its laggard position in the digital era, especially concerning AI. There are references to threats related to the extraction of European data by big tech companies (European Commission, 2020a, 2020b). However, none of these documents analyse in-depth the more active involvement of the states in the US and China and the crucial role of their tech giants.

---

[3] https://ec.europa.eu/commission/presscorner/detail/en/IP_18_4581.

[4] https://www.ft.com/content/f24c844d-18dc-4c8b-a6bf-61526f084f4f?segmentId=b0d7e653-3467-12ab-c0f0-77e4424cdb4c.

[5] https://www.ft.com/content/a169921d-4744-4c16-8ae8-028d52bb655c.

[6] By 2019, Europe had 3.6% of the market capitalization value of the world's 70 largest digital platforms (UNCTAD, 2019).

In a white paper on AI called "A European approach to excellence and trust" the European Commission proposes a set of STI-policy initiatives to strengthen Europe's position. It recommends encouraging venture capital and increased research efforts and skill formation with the ultimate goal "to achieve an 'ecosystem of excellence' along the entire value chain, starting in research and innovation, and to create the right incentives to accelerate the adoption of solutions based on AI, including by small and medium-sized enterprises (SMEs)" (European Commission, 2020b, p. 3). The proposal overlooks the global dominance of US and Chinese tech giants, including their capacity to acquire promising start-ups and to garner intellectual rents from European research institutions and innovating firms. As we showed in Chapter 3, tech giants are in a privileged position to integrate and exploit any newly created European expertise into their Corporate Innovation Systems. This implies that the EU may end up using taxpayers' money to create knowledge and skills absorbed by tech giants.

The proposed AI act is focused on regulating the use of data in connection with AI. It makes a distinction between prohibited, high-risk and normal applications. Examples of prohibited AI are remote biometric identification excepting for searching for potential victims of a crime, terrorist threats or prosecution of perpetrators or suspects of certain crimes. AI systems used to evaluate the credit score or creditworthiness of natural persons and for migration, asylum and border control management are given as examples of high-risk. Within this regulatory frame, the declared objective is to promote AI innovation and a single market for AI applications in Europe. Since there are no explicit limits on access to this market, the formation of the single market will offer opportunities for further expansion primarily for the US tech giants.

It has been argued that the very size of a unified market offers the EU the opportunity to set standards that might then spill over and shape applications worldwide. Another argument for euro-optimism is that new technological developments and especially edge-computing will offer less massive scale advantages and open up for European competitors to the tech giants. The current strategy is rooted in an EU history where free mobility of capital, labour and commodities were established as basic principles and where industrial policy has been defined as competition distortion. The Airbus state support, the single example of ambitious and successful industrial policy at the European level, was triggered by the threat of Boeing establishing a world monopoly. A similar assessment of the position of tech giants would open up for more ambitious industrial policy aiming at creating high-tech public corporations (Archibugi & Mariella, 2021).

8.2.2   The United States

In 2019, there was a break with the permissive antitrust policies that became the rule in the US since the 70s and 80s (Glick, 2019). The US Congress opened an antitrust investigation against Google, Amazon, Facebook and Apple. They were found guilty in 2020 and several prosecutions are on their way. Since then, tech giants have been increasingly put into question, denounced by their use and abuse of private data,[7] anti-competitive practices,[8] and tax avoidance. Among other recent claims, California general attorneys joined the Federal Trade Commission to investigate Amazon for potential abuse of its market power.

It is still to be seen if these attempts will end up altering tech giants' market power. Yet, they represent a turning point from the past and signal that their lobbying may not be as effective as it has been so far. With Google leading the ranking of global lobbying expenditures,[9] tech giants have so far been successful. For instance, in 2015, the US Federal Trade Commission (FTC) mistakenly released portions of a report by its Bureau of Competition staff regarding the Google investigation. The legal staff recommended prosecuting Google. The FTC instead opted to close its investigation after Google committed to change some of its data-driven business practices,[10] something it has not done so far as the US Congress investigation showed (US Congress. Subcommittee on antitrust, commercial and administrative law, 2020).

---

[7] See for instance https://www.ft.com/content/4ade8884-1b40-11ea-97df-cc63de1d7 3f4?segmentId=b0d7e653-3467-12ab-c0f0-77e4424cdb4c.

[8] See for instance https://www.ft.com/content/e56d2820-4cef-11ea-95a0-43d18e c715f5.

[9] https://www.washingtonpost.com/technology/2020/01/22/amazon-facebook-goo gle-lobbying-2019/.

[10] http://graphics.wsj.com/google-ftc-report/.

8.2.3   *China*

The last country to regulate its tech giants' excessive market power was China. Just before Ant Group's IPO in Shanghai, the Chinese government announced antitrust regulations for digital companies forcing Alibaba to cancel the IPO.[11] Claims against Alibaba were only the first step. As we explained in Chapter 6, the Chinese state ordered to terminate Ant Group's financial activities keeping only its e-payments business. Alibaba was also fined USD 2.75 billion by the State Administration of Market Regulation. Further anti-monopoly investigations are on their way against the online food courier Meituan[12] and, as this book is being finished, Tencent is about to receive about USD 1.5 billion fine. Tencent is accused of not properly reporting past acquisitions and for anticompetitive practices in some of its businesses, in particular music streaming.[13]

8.3   ALTERNATIVE FUTURES

8.3.1   *Antitrust and Natural Monopolies*

Capitalism is at a crossroad, even considering that the above-mentioned attempts to regulate tech giants succeed. The latter is still to be seen given—among others—the ongoing battle between Facebook, who recently won an antitrust reprieve,[14] and the US Federal Trade Commission, who responded filing a new suit.[15] Several variables are at stake in the near future in relation to corporate and political powers and involving antitrust, data governance and innovation.

Open questions include whether the US, China and the EU—as well as other countries—advance in a more stringent antitrust that even reverses

---

[11] https://www.ft.com/content/1a4a5001-6411-45fa-967c-0fd71ba9300b.

[12] https://www.bloomberg.com/news/articles/2021-04-26/china-investigates-meituan-for-suspected-monopolistic-practices?cmpid=BBD043021_CN&utm_medium=email&utm_source=newsletter&utm_term=210430&utm_campaign=china.

[13] https://www.reuters.com/world/china/exclusive-china-readies-tencent-penalty-antitrust-crackdown-sources-2021-04-29/.

[14] Retrieved from https://www.ft.com/content/75b74e7a-02d8-430d-b9fa-70a9235b875a?segmentId=b0d7e653-3467-12ab-c0f0-77e4424cdb4c last access July 9, 2021.

[15] Retrieved from https://www.ftc.gov/news-events/press-releases/2021/08/ftc-alleges-facebook-resorted-illegal-buy-or-bury-scheme-crush last access August 20, 2021.

already approved mergers and acquisitions. Or will competition agencies' attempts to regulate big tech be lost in legal procedures, administrative burden and relatively small fines considering tech giants' enormous revenues? Tech giants will argue against antitrust law-making that they are big because they are more efficient. And they may be right since most of the digital services engender natural monopolies and there are serious doubts on whether federated solutions that promote a less asymmetric digital economy will be able to overcome network effects and the quasi-infinite content available on platforms like Google's YouTube or Facebook (O'Neil et al., 2021). Seen from this perspective, there is a need for imagination and boldness when looking for alternatives to governance through either private (but heavily regulated) or public monopolies. Another open question refers to whether these solutions could take place at a global level or whether states' limited power and geopolitical conflicts will, at most, lead to national solutions.

Concerning antitrust policies and tech giants' market dominance, breaking them up remains an option. Still, it would require a clear stand of the US state against ambassadors of its hegemony. Furthermore, considering the tight technological cooperation among data-driven intellectual monopolies (see Chapters 3–5), even if antitrust policies force them to break up, they may still work together, sharing databases and research results to maximize extracted rents. Unlike US tech giants, their Chinese counterparts were compelled to have separate businesses from the start. For instance, Alibaba is formally separated from Ant Group and Alibaba Health. However, since they are all controlled by the same holding company, it remains easy to share datasets to boost each business.

Anyway, the main problem with breaking them up or fostering competition is that tech giants' digital services engender natural monopolies. Think of search engines and how inefficient it would be to have ten places to search instead of one, therefore ten different AI models processing fewer data each than a single one that constantly learns and improves from processing billions of searchers. Since the more searches a deep learning AI model processes the better it gets, clearly centralizing all web searches in one engine would be preferred. Yet, they cannot be at the expense of public access to data and knowledge, including the AI models that result from processing society's data.

This points to fundamental dilemmas. It may be argued that specific fields of digital services, such as search engines and social networking, constitute global natural monopolies. Nonetheless, there are strong arguments for antitrust action because tech giants use their dominant market position to monopolize strategic technologies. One obvious alternative, to "nationalize" the activities and leave it to national or (non-existent) global governments to manage the activities, would raise other types of problems in a world with imperfect democracy and political surveillance. It is a major challenge for progressive political forces to come up with new kinds of solutions to these dilemmas.

8.3.2   The Role of Tech Giants and Industrial Digitalization

Another dimension to be considered when thinking of possible futures concerns what Margrethe Vestager—the EU's Executive vice-president of the European Commission for "A Europe Fit for the Digital Age"— sees as the second phase of digitalization: industrial digitalization, a phase that—she argues—could give the EU some leverage considering its position in major industries.[16] However, this second wave also presents new openings for a further expansion of tech giants beyond their current platforms (see Table 4.2 in Chapter 4). Among others, we are already seeing this in the automobile and healthcare industries.

Autonomous vehicles will expand the time people spend in front of a screen and they offer new markets for digital services. This explains the priority that tech giants give to this field. Alphabet has Waymo, a company dedicated to autonomous driving technology development, and Amazon acquired the self-driving start-up Zoox for over $1.2bn in 2020 and invested in the self-driving start-up Aurora in 2019.[17] In China, Alibaba, Tencent and Baidu cooperate with Audi and with FAW Group Corporation and Volkswagen's joint venture in data analysis, internet-vehicle platform building and intelligent urban transport.[18] Even Foxconn is getting ready to serve tech giants in this industrial shift, expanding

---

[16] Retrieved from https://www.project-syndicate.org/onpoint/eu-regulations-for-the-digital-economy-by-margrethe-vestager-and-anu-bradford-2021-05 last access May 25, 2021.

[17] Retrieved from https://www.ft.com/content/37ae69d9-f160-48c3-b3c5-736730c110ce and https://www.wired.com/story/amazon-aurora-self-driving-investment-funding-series-b/ last access June 4, 2021.

[18] https://www.audi-mediacenter.com/en/press-releases/audi-strengthens-partnerships-with-chinese-tech-giants-6711.

its manufacturing capabilities from smartphones and other devices, to vehicles.[19]

In healthcare digitalization, Alphabet (Google's parent), Apple and Amazon, and to a lesser extent Facebook and Microsoft, are active investors and they engage in major development efforts. Alphabet's approach to healthcare includes Verily Life Sciences, the originally UK start-up DeepMind specialized in AI that includes healthcare projects and Calico that focuses on ageing and age-related diseases. Apple also has its wearable, the Apple Watch, and uses its data in multiple associated healthcare initiatives with universities (Rikap, 2018). Amazon acquired the online pharmacy PillPack in 2018 and Health Navigator in 2019. Amazon Web Services provides specific AI solutions for healthcare and works with universities and hospitals in applying AI to diagnosis, precision medicine, voice-enabled technologies and medical imaging.[20] As we mentioned in Chapter 5, its wellness tracker Halo can even monitor our mood.[21] And Microsoft has just acquired Nuance, a cloud-based system for hospitals and doctors. When the purchase was announced Microsoft's CEO, Satya Nadella, tweeted "AI is technology's most important priority, and healthcare is its most urgent application".

It is still unclear to what extent the digitalization of new industries will result in tech giants conquering these industries and subordinating old leaders. One alternative scenario could be the emergence of a new type of *combined intellectual monopolies* reflecting cooperation between tech giants and intellectual monopolies from other sectors, such as big pharma and leading automotive corporations.

Concerning technological cooperation, Google's Verily Life Sciences partnered with GlaxoSmithKline for a project called Galvani Bioelectronics, and Onduo is a joint venture between Verily and Sanofi (CBInsights, 2018). Yet, big tech companies have the data, the required analytics skills and idle funds to lead other industries' digitalization.

---

[19] https://www.ft.com/content/b229250d-5d9e-4bb1-bb91-e57888233a98.

[20] https://www.beckershospitalreview.com/healthcare-information-technology/pittsburgh-health-data-alliance-partners-with-amazon-web-services-5-things-to-know.html https://www.beckershospitalreview.com/artificial-intelligence/amazon-to-beth-israel-deaconess-tell-us-how-ai-can-make-your-hospital-more-efficient.html.

[21] https://www.theverge.com/2020/8/27/21402493/amazon-halo-band-health-fitness-body-scan-tone-emotion-activity-sleep.

Overall, how and who will set the R&D orientation and harvest associated profits in these industries remains an open question. The future of tech giants' data monopolies and data harvesting will depend on data governance regimes.

8.3.3   Data Privacy Acts or Socializing Data?

Data privacy acts seek to protect the privacy of the individual citizen and limit tech giants' power. Seen in a different perspective they contribute to further knowledge privatization by fostering individual property over data. Concerning data privacy, Hoeyer (2020) opened a discussion on the concept of patient data. He explains that so-called "patient data" are actually data of relations and processes, not exclusive of the patient. Data include information on the treating physician, the hospital or clinic, the laboratory delivering test results, etc. The same could be said of a Facebook post that receives likes and is shared by different people or a purchase that by definition involves two sides (supply and demand). Digital privacy acts are ill-equipped when data points reflect social relations.

An additional question is if those data privacy acts effectively limit data-driven intellectual monopolies from profiting from harvested data. Aho and Duffield (2020) analysed the GDPR and conceptualized it as a policy aimed at protecting individual freedoms. They explain that the GDPR legislation does limit cases like the Cambridge Analytica scandal but does not impede tech giants' data harvesting. Furthermore, there are many loopholes in the regulation. Most US tech giants have their European base in Ireland, whose Data Protection Commission is not complying with Europe's GDPR.[22] Also, concerning data privacy acts, we explained in Chapter 4 that tech giants are developing new approaches to become less reliant on new sources of big data.

Another aspect to consider is that tech giants keep their big datasets secret. Hence, an alternative to fostering data privacy acts could be to follow the advice of several NGOs and advocacy groups that call for socializing data, such as IT for Change (Korjan & Narayan, 2021). Socializing data could intensify technological competition between tech giants. The latter is already happening, as we showed in particular in Chapter 5 for

---

[22] https://www.politico.com/story/2019/04/24/ireland-data-privacy-1270123.

Amazon and Microsoft. Simultaneously, given that tech giants concentrate the most advanced algorithms and digital infrastructure, socializing data would probably favour them more than any other organization since they are better prepared to process and learn from socialized data. In this respect, socializing data value not only among tech giants but for society at large will require granting access to data and also to the algorithms that have been trained with free harvested data. It will also require the development of public digital infrastructure.

8.3.4   On the Future of Techno-Nationalism

Finally, in this exercise of exploring alternative futures, concerning socializing access to data and knowledge (such as algorithms), there is yet another dimension to be considered. Will techno-nationalisms fostered by the pandemic and the US–China conflict be reinforced in the near future? China's commitment to becoming the global AI leader by 2030 and the US AI policy guided by the concern of the US state and its tech giants about China and its own giants point in that direction. However, we have seen a rise in global cooperation in 2021 driven by the OECD and the G20. Although initially mostly focused on advancing towards a global corporate tax structure for large multinationals, this points to possible commitments to reach global agreements. Among them, the G20 is focused on a recovery from the pandemic driven by intangibles, in particular digital goods. Could these instances of negotiation and expressions of will ultimately contribute to tackling major challenges by tilting the scale in favour of common and public access to knowledge and data—including digital infrastructure—or will they end up favouring the privileged position of data-driven intellectual monopolies and their home states?

In this context, exclusively promoting the use and uptake of AI in peripheral countries will expand the market for cloud services and, if not combined with other types of policies, will further reinforce tech giants' current dominance. At the international level, it will reinforce the tendency of knowledge extraction by lead countries from laggards through different mechanisms:

1. National investments in research excellence can, if successful, be extracted through uneven collaboration between research organizations and tech giants.

2. Tech giants and local concentrations of AI competence in the US are attractive employers for elite scholars.
3. National efforts to foster high-tech entrepreneurs could be usurped by tech giants' acquisition strategies.
4. National efforts to establish collaborative innovation may be dominated by foreign tech giants operating in the host country.
5. Speeding up diffusion and broadening fields of application will expand the market of tech giants' services, including cloud services.

Summing up, although the future remains uncertain, most alternative scenarios point to tech giants and their home countries as remaining in dominant positions and as main beneficiaries at the expense of the rest of the world. On this background, we introduce next some policy recommendations aimed at contributing to create a less polarized future, where collectively solving major societal, ecological and health challenges drive production and policymaking.

8.4   Policy Recommendations

In a world characterized by a high degree of uncertainty, there are no magic bullets. We think of policy recommendations in the context of broader social and political transformations, fostered by public agency and the acknowledgement that the future, if left alone, would be worse. We agree with Prainsack (2020) who criticizes that data policies have been excessively concentrated on data protection and ownership leaving questions on the distribution of power aside. She suggests considering who benefits, at what cost and for what purposes. Our policy recommendations move along those directions.

We add to this a concern about the impact of tech giants' intellectual monopoly status on innovation both in terms of its rate and direction. Since AI has great potential to transform many aspects of life as well as all kinds of economic activities, who remains in charge of shaping it is of crucial importance. Current AI applications reflect the priorities of tech giants, mostly about entertainment in private consumption and surveillance solutions for the military and police departments, which can be more easily monetized, and little about the environment and helping the marginalized regions and people. Hence, the detrimental effects of the dominant influence of tech giants on technological trajectories go beyond limiting access to—and therefore curtailing the development of—knowledge and innovation. As far as they remain as agenda setters, the rate and direction of AI innovation will not be put at the service of societies' major challenges.

This book has focused on the monopolization of data and knowledge, which engenders a specific type of monopoly power whose effects go beyond textbook monopolies (higher price, restricted supply and misallocation). Broader perspectives on economic power should be considered to fully tackle the effects of (data-driven) intellectual monopolies. Intellectual monopolies can limit knowledge spillovers beyond the markets where they operate. Furthermore, in the case of data-driven intellectual monopolies, there are long-term consequences of stifling innovation at the (global) system level. These companies can offer free products or low prices to consumers for the sake of harvesting more data that will result in a further concentration of digital intelligence capturing higher intellectual rents, thus appropriating value from society, typically when exchanging with other firms.

A step towards an effective regulation is to take seriously that we are dealing with digital global public goods. Digital platforms are globally produced by society at large and since digital services—in particular those in the hands of big tech companies—tend to be natural monopolies, they should be envisioned as global public goods. While the feasibility of developing innovative forms of governance that correspond to this perspective depends, among others, on public pressure, other redistributive measures could be taken into consideration in the meantime.

G20 has recently suggested the taxation of profits based on the location of customers. The OECD and the G20 have also been advocating for a global minimum corporate tax rate to tackle base erosion and profit shifting (OECD/G20 Inclusive Framework on BEPS, 2020). However, the so far agreed minimum tax rate seems insufficient to tackle the effects of tech giants. This minimum tax rate will only be imposed if companies earn more than a 10% profit rate, which would not affect Amazon's earnings. This company has developed an internal transfer system where losses from outside the US and Europe are concentrated in Luxemburg together with the company's EU profits. As a result, it ends up receiving tax credit in Europe, globally paying little or no tax at all (Phillips et al., 2021).

Moreover, the way profits are calculated, considering R&D as an expense and not as an investment, contributes to providing artificially reduced profit figures for intangible-intensive firms. This is all the more true for tech giants; they use intangible assets deferrals—often dubiously calculated—to inflate their costs, thus reducing declared profits. Overall, tech giants and other leading corporations take advantage of accounting rules and use financial instruments to alter their accounting data to pay lower taxes.

Therefore, reforms should aim at taxing digital companies' *revenues* rather than profits. Amazon is an extreme case illustrating the limits of taxing profit. It is a company that barely declares profits but whose market capitalization has risen about 1789% in the last decade. This points to yet another option: to tax shareholders in line with the rise of corporations' market capitalization. Limiting buybacks and other mechanisms that artificially increase shares' prices should also be put in place. Apple has been particularly active in terms of buybacks.

Moreover, data-driven intellectual monopolies earn rents through mechanisms that may be paralleled to land rents usurped by landowners where their source of rents is access to data. This parallel points to a progressive tax on owners of digital databases created from centralizing third-party—individuals or organizations—data. Tech giants should then pay according to the size of their databases. A relatively simple way to proxy size (if required information is not provided) is to consider datacentres' square metres.

Unlike land, these taxes should not be charged according to where the data centres are located. This would benefit core countries, in particular the US, home to 40% of the world's hyperscale data centres (Synergy Research Group, 2019). The tax collection could take place at the global level considering the global scale of these corporations, but also to assure that no country sets a zero tax on data centres, therefore favouring their relocation at the expense of the rest of the world. Tax revenues could be shared according to countries' share of Internet penetration, or—even better for development—through a progressive scheme favouring less developed countries.

Besides addressing the effects of data concentration and reaping back the rents appropriated by tech giants, another necessary reform refers to the global intellectual property regime. The current regime favours knowledge owners but seldom benefit actual inventors (Dreyfuss & Frankel, 2014; Pistor, 2019). It also limits global learning and stifles innovation since knowledge is a cumulative process where society builds new knowledge based on previous knowledge. If the latter is only accessible by paying an intellectual rent, promising knowledge avenues may be discarded to avoid such costs. Alternative and less harmful methods to foster innovation need to be promoted, including prizes financed with public money.

8.4.1 *Focusing on Underdeveloped Countries*

While all these policies certainly require global enforcement to effectively tilt the scale against data-driven intellectual monopolies, non-core countries should not wait for the US, the EU and China to regulate big tech. Peripheral countries are net providers of raw data that are freely harvested and monetized by a handful of data-driven intellectual monopolies mostly from the US and China. This form of data or digital colonialism reinforces underdevelopment (Couldry & Mejias, 2019a, 2019b; Kwet, 2019). Data privacy laws, inspired by the law passed in Europe, are being implemented in countries like Brazil, Thailand, South Korea, South Africa and India.[23] Nonetheless, as we argued above, these regulations are tricky because as much as they try to limit tech giants' access to data, they still fall short and further promote a culture that celebrates intangibles as private property.

Instead, these countries could require data-driven intellectual monopolies to transfer the technology necessary to build data centres using state-owned infrastructure. A step further could be to charge them for privately appropriating data and also force them to store them in local data centres. Some peripheral regions host local data-driven intellectual monopolies, like MercadoLibre in Latin America. Given the importance of some digital services, like e-commerce and e-payments, states should regulate these giants with the same regulations applied to banks because, besides e-payments, e-commerce is intertwined with new forms of private credit and asset management (Frost et al., 2019).

Some peripheral economies are also advancing regulations to reap back part of the profits made by tech giants in their territories. In Indonesia, the state has threatened foreign internet firms to block their services in the country if they do not obtain "permanent establishment" status and, thus, pay Indonesian taxes.[24] Singapore and Malaysia established digital tax regimes for overseas companies in 2020, and Thailand introduced a 7% value-added tax on sales for digital platforms that lack a local subsidiary company and make more than $57,000 a year. A bill was introduced in

---

[23] https://www.endpointprotector.com/blog/data-protection-legislation-around-the-world-in-2020/.

[24] https://www.reuters.com/article/us-indonesia-tax-internet-idUSKCN0W20QM.

May 2020 in the Philippines to raise funds to face pandemic-associated expenses. It taxes big providers such as Facebook, Google, Netflix and Spotify.[25]

In countries where tech giants' operations are not yet widely developed, there is a chance for national and even local authorities to develop state-owned or cooperative platforms while limiting US and Chinese tech giants' access. Tech giants are now expanding their cloud business to the peripheries. This is the case of cloud computing in Southeast Asia.[26] There is, therefore, a small window of opportunity for states to coordinate and offer state-owned or cooperative public clouds before data-driven intellectual monopolies capture this market.

In Indonesia, a state-owned enterprise already manages the digital payment service LinkAja.[27] Brazil has seen the emergence of several public platforms since the pandemic, like the delivery platform FiqueNoLar that operates in the northern part of the country, where the most popular private delivery apps did not have a service.

In fact, although their chances to succeed in triggering network effects will be smaller, state-owned or cooperative platforms should also be created in countries where local or global tech giants have an established business. For instance, in October 2020 Argentina announced the creation of an online marketplace called "Correo Compras" to be run by a state-owned company, Correo Argentino, which is also the country's official postal service. And, amid the Covid-19 pandemic, some advocated that the US government should nationalize Amazon and use its logistics network to assure the delivery of essential goods to all US citizens.[28] For these initiatives to succeed and overcome part of the shortfalls of private companies, digital security must be guaranteed and surveillance capitalism overcome, which certainly requires civil society bodies that oversee how data are used and stored.

* * *

[25] https://asia.nikkei.com/Economy/From-Thailand-to-Indonesia-taxes-tighten-for-digital-businesses.

[26] Retrieved from https://www.ft.com/content/1e2b9cd9-f82e-4d3b-a2d8-f20c08bdc3aa?segmentId=b0d7e653-3467-12ab-c0f0-77e4424cdb4c.

[27] https://www.linkaja.id/.

[28] https://inthesetimes.com/article/supply-chain-crisis-nationalize-amazon-coronavirus-covid-19.

Overall, both at the global level and for peripheral countries, a comprehensive plan should be put in place to tackle intellectual monopolies and their effects. It should address the effects of knowledge privatization and assetization, opening space for a new global knowledge regime based on common and public knowledge. Yet, this type of initiative will require significant public pressure, such as what we are currently witnessing for the case of Covid-19 vaccine waiver.

8.5   We Need More Activism

The rise of US and Chinese big tech as the world's most powerful corporations has deepened inequalities at multiple levels: among firms, among workers, among countries and between capitalists and workers. In this context, social disrupts have been an unsurprising recurring outcome, and we have seen them everywhere in the twenty-first century. Even during the pandemic, revolts surged against different forms of inequality in, among others, Chile, Colombia, India and the US.

Demonstrations are increasingly being organized online. The same technology that underlies tech giants' power, used for—corporate and state—surveillance, for broadcasting extreme right and even fascist ideas and that fuels the US–China conflict, is also being used as a counterbalancing weapon. Internet, particularly social networks, is a powerful tool for the organization of grassroots movements. Workers' unions can also learn from each other's experiences and get in contact online. During the pandemic, digital technologies have also been used to organize assemblies, academic conferences, political meetings and so on.

Unions are adapting and workers organizing themselves, and this should be fostered at every possible level. In 2018, Google employees managed to stop the company from renewing an AI contract with the Pentagon and to cancel its plans for a censored search engine for China. And, in 2020, 2000 employees urged the company to cease selling technology to the US police after George Floyd's killing. Google's employees created their union and Amazon warehouse workers in different parts of the world are also organizing and trying to become unionized workers. Even if they fail several times before succeeding, the threat of unionization is already impacting Amazon's working conditions for the good.

These initiatives could inspire workers from other companies. Unions should be reconceived as political actors capable of exercising their influence beyond wage claims. Workers' organization is indispensable to counterbalance the power of intellectual monopolies, given both their global reach and states' internal contradictions and limitations.

8.6   FINAL REMARKS

In this book, we have focused on how countries and corporations compete to establish leadership in AI. It is obvious that those governing the US and China now see leadership in this technology as being of fundamental importance for leadership in world affairs. There is a growing understanding in the rest of the world that AI is key for economic growth and to tackle global challenges related to health, inequalities, climate change and other major ecological challenges. Hence, it is highly problematic that the capacity to lead AI development and enjoy associated profits has become concentrated in a handful of tech giants with roots in the US and China. In the first phase of the ICT revolution, there was a similar concentration of the semiconductor and computer industry in Silicon Valley. That did not hinder other developed regions to use microelectronics to develop new products and more efficient processes supporting their competitiveness.

Is this time different? We think that it is for a few interconnected reasons. Scale economies are crucial and they operate at different levels. As indicated, some digital services operate in what seems to be global natural monopolies and, at the same time, incumbents have the major dynamic advantage of digital learning where the technological lead increases with the scale of accessed data. While it may be argued that microelectronics was a general-purpose technology, manufacturing was its primary field of application, whereas AI can be applied in all sectors and not least in those crucial for human welfare, such as education and health. What's more, it may revolutionize the innovation process itself.

This is why the current geopolitical distribution of corporate capabilities in AI is unsustainable and undefendable. As indicated in the introduction of this chapter, there are two major alternatives to the current situation. The first is that the techno-nationalist strategies of the US and China are reproduced in other parts of the world. In the case of the EU, it would require both a stronger federation and a break with historical bias in favour of open markets. Since the size of the market determines the volume of data, it would require a movement towards regional integration in other parts of the world, such as in Africa and Latin America. In these cases, a pragmatic approach to the current conflict between the US and China may be useful, since the very conflict may offer access to critical elements of knowledge. However, this alternative points to new barriers to access to knowledge and information and an escalation of conflicts with a major risk of military confrontations.

The other alternative is that the rest of the world forces the US and China to re-establish a collaboration to develop and use AI focused on solving global problems including poverty and economic underdevelopment. It would require a new global regime where access to data from abroad is compensated by sharing crucial knowledge including algorithms. To move in this direction, we see the need for a new—truly democratic—Enlightenment. This time, all humans around the world should be given insights into the fundamentals and contradictions of the emerging digital society.

Current stereotypes where the US is seen as a liberal market society and China as completely dominated and managed by the state and the Communist Party are misleading and dangerous. And they point to irrelevant alternative modes of governance. Recent developments have made it abundantly clear that even these two extremes are mixed societies. In both countries, the state is highly active in steering technologies and there are private interests that go against government regulations. Without a new discourse that goes beyond the duality between state and market and between private and public, it is difficult to see how to realize the second alternative. New forms of citizen engagements in building global cooperative endeavours will be elements of a third alternative. It is difficult to imagine such a third alterantive without new forms of transnational citizen activism resulting in new forms of global cooperation.

AI evokes both dystopian and utopian perspectives. Its breakthrough should engage citizens all around the world to use their imagination to think about how to construct a fair and sustainable society for all, where global governance plays a stronger role and where knowledge is widely shared within and across nation states.

References

Aho, B., & Duffield, R. (2020). Beyond surveillance capitalism: Privacy, regulation and big data in Europe and China. *Economy and Society, 49*(2), 187–212.

Archibugi, D., & Mariella, V. (2021). Is a European recovery possible without high-tech public corporations? *Intereconomics, 56*(3), 160–166.

CBInsights. (2018). *How Google plans to use AI to reinvent the $3 trillion US healthcare industry.*

Couldry, N., & Mejias, U. A. (2019a). Data colonialism: Rethinking big data's relation to the contemporary subject. *Television & New Media, 20*(4), 336–349.

Couldry, N., & Mejias, U. A. (2019b). *The costs of connection: How data is colonizing human life and appropriating it for capitalism.* Stanford University Press.

Dreyfuss, R., & Frankel, S. (2014). From incentive to commodity to asset: How international law is reconceptualizing intellectual property. *Michigan Journal of International Law, 36*(4), 557–602.

European Commission. (2017). *CASE AT.39740. Google search (shopping)* [Antitrust Procedure]. European Commission.

European Commission. (2020a). *Communication from the commission to the European parliament, the council, the European Economic and social committee and the Committee of the Regions: A European strategy for data.* European Commission.

European Commission. (2020b). *White paper on Artificial Intelligence—A European approach to excellence and trust.* European Commission.

Frost, J., Gambacorta, L., Huang, Y., Shin, H. S., & Zbinden, P. (2019). *BigTech and the changing structure of financial intermediation.* Bank of International Settlements.

Glick, M. (2019). Antitrust and economic history: The historic failure of the Chicago School of Antitrust. *The Antitrust Bulletin, 64*(3), 295–340.

Hoeyer, K. (2020). Data promiscuity: How the public–private distinction shaped digital data infrastructures and notions of privacy. *Humanities and Social Sciences Communications, 7*(1), 1–8.

Korjan, A., & Narayan, V. (2021). *Socializing data value: Reflections on the state of play.* IT for Change.

Kwet, M. (2019). Digital colonialism: US empire and the new imperialism in the global south. *Race & Class, 60*(4), 3–26.

OECD/G20 Inclusive Framework on BEPS. (2020). *Addressing the tax challenges arising from the digitalisation of the economy.* OECD.

O'Neil, M., Cai, X., Muselli, L., Pailler, F., & Zacchiroli, S. (2021). *The coproduction of open source software by volunteers and big tech firms.*

Phillips, R., Pyle, J., & Palan, R. (2021). *The Amazon method: How to take advantage of the international system to avoid tax* [City PERC policy report]. University of London.

Pistor, K. (2019). *The code of capital: How the law creates wealth and inequality.* Princeton University Press.

Prainsack, B. (2020). The value of healthcare data: To nudge, or not? *Policy Studies*, 1–16.

PWC. (2020). *Global Top 100 companies by market capitalisation.* United States.

Rikap, C. (2018). Innovation as economic power in global value chains. *Revue D'économie Industrielle, 163*, 35–75.

Synergy Research Group. (2019). *Hyperscale data center count jumps to 430; Another 132 in the pipeline.* https://www.srgresearch.com/articles/hyperscale-data-center-count-jumps-430-mark-us-still-accounts-40

UNCTAD. (2019). *Digital economy report 2019: Value creation and capture–Implications for developing countries.* United Nations.

US Congress. Subcommittee on antitrust, commercial and administrative law. (2020). *Investigation of Competition in digital markets.* US Congress.

INDEX

The manufacturer's authorised representative in the EU is Springer
Nature Customer Service Centre GmbH, Europaplatz 3, 69115 Heidelberg,
Germany. If you have any concerns regarding our products, please
contact ProductSafety@springernature.com

Printed and bound by CPI Group (UK) Ltd, Croydon, CR0 4YY
29/04/2026
02099478-0002